An Introduction to Cyber Modeling and Simulation

Wiley Series in Modeling and Simulation

The Wiley Series in Modeling and Simulation provides an interdisciplinary and global approach to the numerous real-world applications of modeling and simulation (M&S) that are vital to business professionals, researchers, policymakers, program managers, and academics alike. Written by recognized international experts in the field, the books present the best practices in the applications of M&S as well as bridge the gap between innovative and scientifically sound approaches to solving real-world problems and the underlying technical language of M&S research. The series successfully expands the way readers view and approach problem solving in addition to the design, implementation, and evaluation of interventions to change behavior. Featuring broad coverage of theory, concepts, and approaches along with clear, intuitive, and insightful illustrations of the applications, the Series contains books within five main topical areas: Public and Population Health; Training and Education; Operations Research, Logistics, Supply Chains, and Transportation; Homeland Security, Emergency Management, and Risk Analysis; and Interoperability, Composability, and Formalism.

Spatial Agent-Based Simulation Modeling in Public Health: Design, Implementation, and Applications for Malaria Epidemiology
by *S. M. Niaz Arifin, Gregory R. Madey, Frank H. Collins*
The Digital Patient: Advancing Healthcare, Research, and Education
by *C. D. Combs (Editor), John A. Sokolowski (Editor), Catherine M. Banks (Editor)*

An Introduction to Cyber Modeling and Simulation

Jerry M. Couretas

WILEY

This edition first published 2019

Registered Office
John Wiley & Sons, Inc., 111 River Street, Hoboken, NJ 07030, USA

Editorial Office
111 River Street, Hoboken, NJ 07030, USA

For details of our global editorial offices, customer services, and more information about Wiley products visit us at www.wiley.com.

Wiley also publishes its books in a variety of electronic formats and by print-on-demand. Some content that appears in standard print versions of this book may not be available in other formats.

Library of Congress Cataloging-in-Publication Data

Names: Couretas, Jerry M., 1966– author.
Title: An introduction to cyber modeling and simulation / Jerry M. Couretas.
Description: Hoboken, NJ : John Wiley & Sons, 2019. | Series: Wiley series in modeling and simulation | Includes bibliographical references and index. |
Identifiers: LCCN 2018023900 (print) | LCCN 2018036433 (ebook) | ISBN 9781119420811 (Adobe PDF) | ISBN 9781119420835 (ePub) | ISBN 9781119420873 (hardcover)
Subjects: LCSH: Computer simulation. | Computer security. | Cyberinfrastructure.
Classification: LCC QA76.9.C65 (ebook) | LCC QA76.9.C65 C694 2018 (print) | DDC 005.8–dc23
LC record available at https://lccn.loc.gov/2018023900

Cover Design: Wiley
Cover Image: © MimaCZ/Getty Images

Set in 10/12pt Warnock by SPi Global, Pondicherry, India

Printed in the United States of America.

V10004604_091418

Dedication

This book is dedicated to Monica, Sophie, and Ella, for the time and patience that they provided to bring this work to fruition. I would also like to thank Jorge and Aida Carpio, for their support and mentoring. Finally, to my parents, Gus and Mary, for providing an example of persistence and faith.

Professionally, I would like to thank Mr. Rick Stotts for introducing me to modern cyber, Professor Bernard Zeigler, for his ongoing support from my student days to the present.

Contents

1

Brief Review of Cyber Incidents

When it comes to national security, I think this [i.e., cyber warfare] represents the battleground for the future. I've often said that I think the potential for the next Pearl Harbor could very well be a cyber-attack. If you have a cyber-attack that brings down our power grid system, brings down our financial systems, brings down our government systems, you could paralyze this country.[1]

Leon Panetta

The 1988 Morris Worm, designed to estimate the size of the Internet, caused the shutting down of thousands of machines and resulted in the Defense Advanced Research Projects Agency (DARPA) funding the first Computer Emergency Response Team (CERT) at Carnegie Mellon University (CMU). As shown in Table 1.1, cyberattacks have continued since 1988, with effects that range from data collection to controlling critical infrastructure.

Table 1.1 also provides a mix of documented cyber incidents, with only the Morris Worm in question, as to malevolent intent. Due to the multiple actors and actions, involving cyberattacks, a conversation around "resilience" (e.g. NIST Cybersecurity Framework) provides a means for communicating about how the overall system will continue to perform, in the face of adversity. In addition, resilience frames the discussion about an organization's operational risk; due to cyber, in this case. More specifically, the resilience view provides a means to organize the challenges associated with measuring and quantifying the broad scope of an organization's cyberattack surface by:

1) Recognizing that the autonomy and efficiencies that information systems provide are too valuable to forego, even if the underlying technologies provide a potential threat to business operations.

---

1 "Cybersecurity 'battleground of the future,'" *United Press International*, 10 February 2011, available at http://www.upi.com/Top_News/US/2011/02/10/Cybersecurity-battleground-of-thefuture/UPI-62911297371939/, accessed on 10 January 2012.

An Introduction to Cyber Modeling and Simulation, First Edition. Jerry M. Couretas.

Table 1.1 Select cyber incidents.

Year	Cyberattack	Objective	Effects
1988	Morris Worm	Understand the number of hosts connected to the Internet	Removed thousands of computers from operation
2003	Slammer Worm	Denial of service	Disabled Ohio's Davis–Besse nuclear power plant safety monitoring system for nearly 5 h
2008	Conficker	Implant malware on target machines	Control target machines
2010	STUXNET	Take control of Siemens industrial control systems (ICS')	Destroyed centrifuges used for Iranian nuclear program
2012	Saudi Aramco (oil provider) business systems (aka Al Shamoon)	Wipe disks on Microsoft Windows-based systems	Destroyed ARAMCO business systems to cause financial losses due to their inability to bill customers for oil shipments
2013	South Korean Banks	"DarkSeoul" virus used to deny service and destroy data	Destroyed hard drives of selected business systems
	US Banks	Distributed Denial of Service (DDoS)	Caused financial losses through banks' inability to serve customers
	Rye Dam (NY)	Access control gates for opening and closing at will	Controlled dam gate system
2014	Sony Pictures	Data breach	Downloaded a large amount of data and posted it on the Internet; 3 wk before the release of a satirical film about North Korea
2015	Office of Personnel Management (OPM) breach	Gain access to information on US Government Personnel	Downloaded over 21 million US Government and contractor personnel files
2017	Equifax breach	Gain access to consumer credit information	Downloaded credit history and private information on over 143 million consumers

2) Understanding that cyber "security" (i.e. the ability to provide an effective deterrent to cyberattacks) is not achievable for most organizations in the short term, so resilience is one way to develop organizational policies and processes around
   a) mitigation scenarios for general cyberattacks
   b) comparing tacitly accepted cyber risk to business risks that we already transfer (e.g. hurricanes, earthquakes, natural disasters, etc.) to other organizations (e.g. insurance companies).
3) Coordinating the challenges associated with an organization's people being a key source of cyber vulnerability.

Resilience, therefore, provides an overarching approach, with some elements already modeled, for bundling the exposure associated with cyber and moving the discussion to a more manageable set of risks; analogous to operational challenges already mitigated or transferred through an organization's policies and processes. In addition, cyber risk management requires analytical evaluation and testable scenarios that enable contingency planning for each respective organization. Cyber risk assessment is a growing area of interest, and an inspiration for developing cyber modeling and simulation techniques.

1.1 Cyber's Emergence as an Issue

The issue of cyber security, somewhat slow to be recognized during information technology's rapid rate of development and dissemination into business enterprises over the last half century, often gets the same level of news coverage as natural disasters or stock market anomalies. While an Office of Personnel Management (OPM)[2] breach disclosing the private information of millions of US civil servants gets a few days of news, a new iPhone release will create weeks of chatter on social networks. Cyber insecurity is much less interesting to the general public than the Internet's entertainment and socialization prospects.

The same market growth for personal computing technologies, however, adds to unforeseen security challenges that networked technologies provide. Increased connectivity, often leading to tighter coupling (i.e. both technically and socially), challenges "open" information system architectures and their intended benefit. In addition, this increased connectivity provides, for the first time, an artificial domain, or space, through which nefarious actors can exercise potentially catastrophic effects. Cyber's ability to deny or manipulate entire regions of a state, at time constants much shorter than current management structures can handle, is a relatively recent realization. For example,

2 https://www.wired.com/2016/10/inside-cyberattack-shocked-us-government/

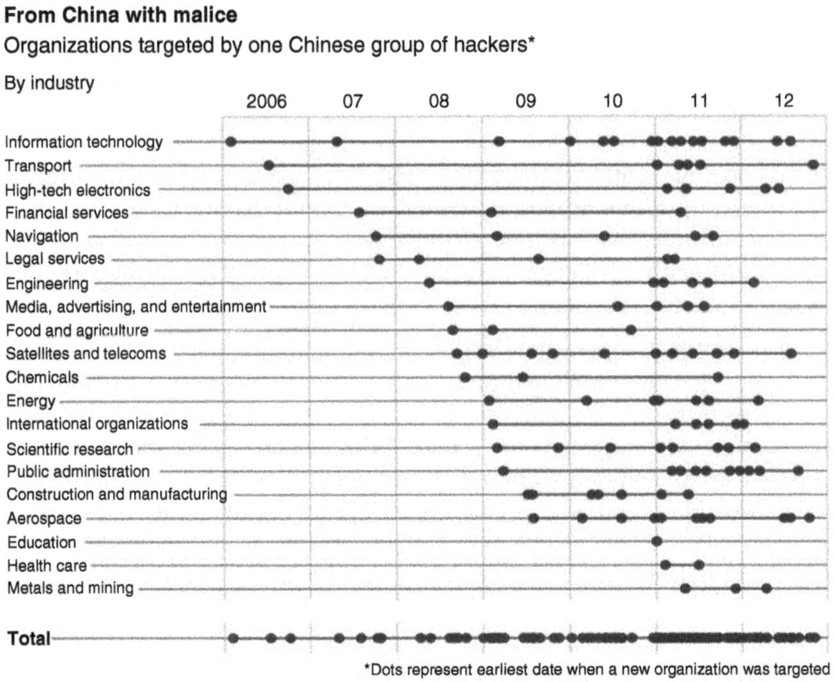

Figure 1.1 Organizations targeted by China. *Source:* Mandiant (2014), Fireeye https://www.fireeye.com/content/dam/fireeye-www/services/pdfs/mandiant-apt1-report.pdf.

by 2015, reports (Frankel et al. 2015; Maynard and Beecroft 2015) on the potentially catastrophic nature of a cyberattack started to emerge. Along with the increasing importance of cyber, as a physical threat, there is an increased awareness, via news coverage (Figure 1.1).

In addition to Figure 1.1's profile of commercial cyber activity, military applications are expanding as well, with notable uses in Estonia and Georgia over the last decade.

1.2 Estonia and Georgia – Militarization of Cyber

For three weeks in 2007, the Republic of Estonia suffered a crippling cyberattack that left government, political, and economic facets of the country helpless (Yap 2009) (Figure 1.2).

This scenario provides a template to examine the policy, training, and technology options of a cyber-attacked state. Estonia's policy options were limited for a number of reasons, including:

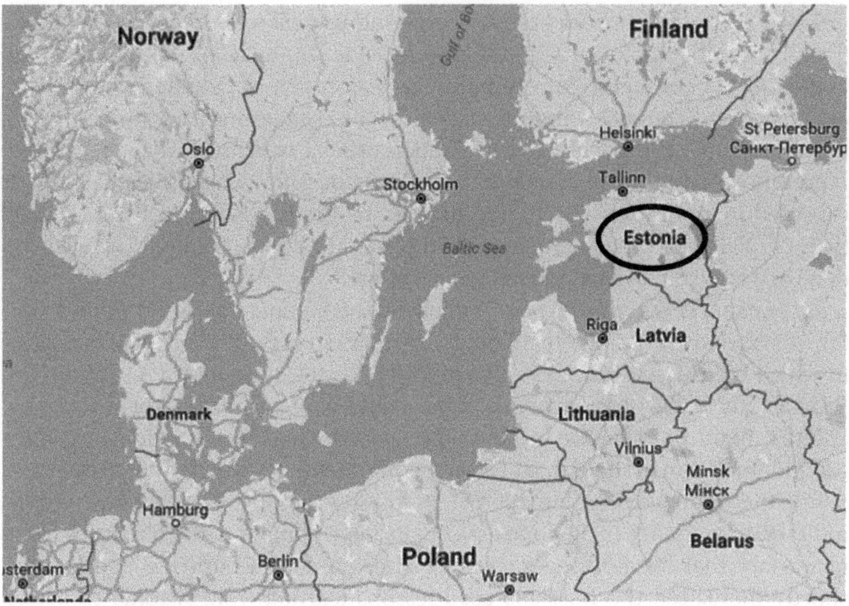

Figure 1.2 Map of N. Europe with Estonia (Google Maps).

- difficulty of attribution
- lack of international standards
- current political environment

Ultimately, unless a cyberattack causes indisputable damage, loss of human life, and can be traced back to a source with high certainty, it is unlikely that a state will conventionally respond in self-defense. Currently, there are no clear international laws,[3] or rules of engagement, that govern the rights of any sovereign state in the event of a cyberattack, without people dying or significant physical damage. The current approach is to take the existing laws and treaties and interpret them to fit cyber domain activities.

However, unlike a conventional attack, there are many more factors that blur the line in cyberspace. Attribution is usually spread across different sovereign states with limited physical evidence. Without a common, and agreed upon, definition of what constitutes a cyberattack, how can nations defend themselves without risking ethical, legal, and moral obligations? The fundamental dilemma a state faces is to balance its retaliatory options with the requisite legal justifications, if they cannot be confident of the source for the attack.

3 The Tallinn Manual (https://ccdcoe.org/tallinn-manual.html) provides one approach for adapting laws of war to cyberspace.

While policy frameworks are still evolving to deal with cyber as a conflict domain, newly employed technologies provide unprecedented platforms for launching cyberattacks. For example, the major part of Estonia's assault suddenly stopped roughly a month after it began, suggesting that a botnet had been leased for the attacks. One Estonian official concluded that the attacks represented "a new form of public–private partnership" where the attacks were executed by organized crime, but directed by the Kremlin. "In Estonia," said then US National Security Agency chief General Keith Alexander, "all of a sudden we went from cybercrime to cyberwarfare."[4]

Some experts (Krepinevich 2012) believe the Estonia attack provided a way for Moscow to test its new technology, cyber weaponry, as a "proof of concept," in which the Russian Business Network (RBN) was given a target to show the Russian authorities how valuable cyber could be. In this way, the attacks on Estonia might be compared to how the Spanish Civil War provided a testing ground for German, Italian, and Soviet equipment and war-fighting concepts. While the evidence is circumstantial, it appears that just as Germany used its military's experience in Spain to assist in its development of the blitzkrieg form of warfare that it employed against Poland, the Low Countries, and France, shortly thereafter, the Russians used lessons learned from Estonia to better integrate cyber operations with traditional military operations in Georgia.

A year after the Estonia attacks, Georgia suffered the world's first mixed cyber–conventional attacks (Beidleman 2009). The cyberattacks were staged to kick off shortly before the initial Russian airstrikes as part of the Russian invasion in August 2008. The cyberattacks focused on government websites, with media, communications, banking, and transportation companies also targeted.

These botnet-driven DDoS attacks were accompanied by a cyber blockade that rerouted all Georgian Internet traffic through Russia and blocked electronic traffic in and out of Georgia. The impact of the cyberattacks on Georgia was significant, but less severe than the Estonia attacks since Georgia is a much less-advanced Internet society. Nonetheless, the attacks severely limited Georgia's ability to communicate its message to the world and its own people, and to shape international perception while fighting the war.

1.3 Conclusions

Modeling the broadly scoped set of systems that "cyber" currently covers, along with their associated effects, is a challenge without specifying the technical, process, or policy aspects of a scenario in question. While constructive modeling

4 Keith B. Alexander, statement before the House Armed Services Committee, 23 September 2010, p. 4.

and simulation has made great contributions to describing the technical aspects of engineered systems for their testing and development, murky process and policy threads are still very much present in most cyber case studies – often providing the real security issues for the systems at risk. For example, computer technologies are often, simply, the implementation of processes for complex systems that support us. A "cyber" attack is really an attack on one of these processes we trust for our day-to-day business.

Cyber's overarching use has implications across both a country's business systems and its supporting civil infrastructure. Understanding the current state, in the cyber domain, therefore requires accurately assessing our systems and evaluating their maturity from a cyber standpoint. Using these assessments for defensive, or resiliency, analysis is the first step to verify M&S for cyber systems. Real-world cyber scenarios then use these assessments, as baselines, to represent both the scope and scale of networks with technologies and configurations that can easily span multiple generations of information technology.

2

Cyber Security – An Introduction to Assessment and Maturity Frameworks

There are security implications that result from our incorporating computer automation, or cyber, into business systems and industrial control systems that underpin almost everything we do. Assessing these cyber systems, to ensure resilience, is performed through a number of well-known frameworks to develop an initial understanding, or baseline, of our current system security levels.

Assessments often begin with an asset prioritization, a "Crown Jewels Analysis[1]" (MITRE) being one example, with more detailed evaluations developed from this initial structure. Figure 2.1 provides an example "Enterprise Risk Analysis" structuring designed to perform this high-level prioritization, with detailed process modeling showing system dependencies for structural evaluation. Component-level assessment, or penetration testing, is then used at the technology level to inventory the system's architecture.

As shown in Figure 2.1, network evaluation spans from an overall key asset prioritization to specific network components. This can include using dependency or attack graphs, during process modeling, to highlight specific scenarios.

2.1 Assessment Frameworks

The standard Confidentiality, Integrity, and Availability (CIA) information security triad, complimented by subject matter expertise and proven use cases, provides a foundation for computer defense operations.

2.2 NIST 800 Risk Framework

The standard NIST SP 800-30 is used for baseline cyber system security evaluation (Table 2.1).

1 https://www.mitre.org/publications/systems-engineering-guide/enterprise-engineering/systems-engineering-for-mission-assurance/crown-jewels-analysis

An Introduction to Cyber Modeling and Simulation, First Edition. Jerry M. Couretas.

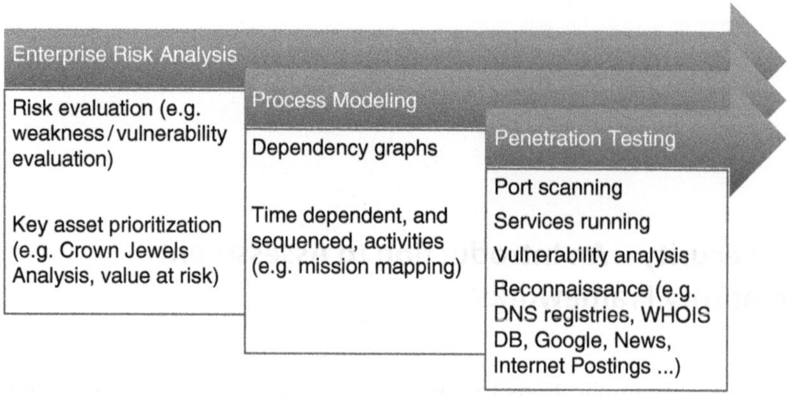

Figure 2.1 Assessment levels – enterprise risk, process modeling, and vulnerability analysis.

Table 2.1 NIST SP 800-30 risk assessment.

System charaterization
Threat identification
Vulnerability identification
Control analysis
Likelihood determination
Impact analysis
Risk determination
Control recommendations
Documentation

Table 2.1's risk descriptions attempt to capture all of the issues and inconveniences that characterize the precursors of a training event for cyber planning. These are items that help with "designing in" cyber-protective measures during the development process. For example, there are two primary ways (Gelbstein 2013) to look at current information enterprise analysis:

1) **Business Impact Analysis** (BIA) – used for incident response, BIA is quantifiable, relies on credible data, and drives crisis management, including disaster recovery and business continuity:
   a) MITRE Crown Jewels Analysis
2) **Enterprise Risk Management** (ERM) – used for incident forecasting is a challenge to quantify, relies on assumptions and drives mitigation plans, the risk register and the return on security investment (ROSI) calculation:
   a) NIST Cybersecurity Framework (CSF)
   b) Risk Management Framework (RMF)

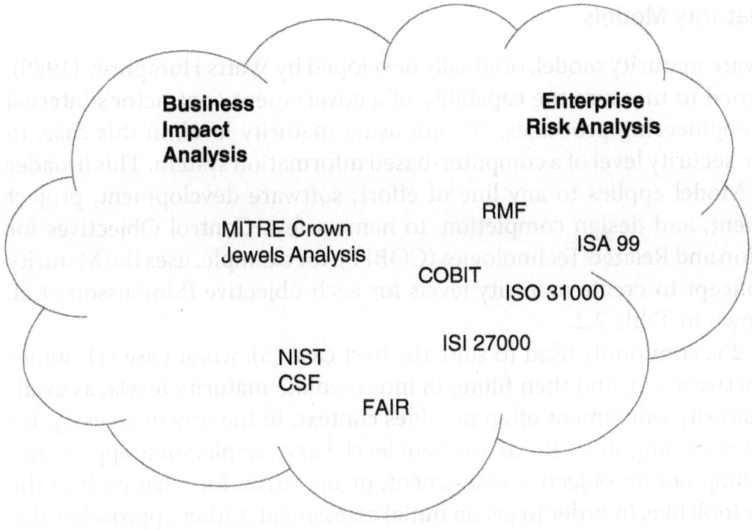

Figure 2.2 Information security – business impact and enterprise risk analysis.

c) ISO 31000
d) ISI 27000 series

Figure 2.2 shows how assessment and management approaches are used to help define business impact and enterprise risk analysis categories for information system evaluation.

As shown in Figure 2.2, information security assessment can take the form of business impact or enterprise risk analysis. In addition, taxonomy information is provided by both the Factor Analysis of Information Risk (FAIR) model (Jones 2005) and the NIST CSF, in further defining the terms used in implementing relevant security policy. Each of the frameworks is designed to improve system security, in the familiar resilience steps[2] of:

- Identify
- Prevent
- Detect
- Respond
- Reconstitute

Using the "conceptual model" provided by the standard steps of resilience provides the modeler with an approach for modeling cyber defense processes. In addition, this "conceptual model" compliments the standard maturity model assessment.

2 https://www.us-cert.gov/ccubedvp/cybersecurity-framework

2.2.1 Maturity Models

The software maturity model, originally developed by Watts Humphrey (1989), was designed to measure the capability of a government contractor's internal software engineering processes. We are using maturity level, in this case, to look at the security level of a computer-based information system. This broader Maturity Model applies to any line of effort; software development, project management, and design completion, to name a few. Control Objectives for Information and Related Technologies (COBIT), for example, uses the Maturity Model concept to create maturity levels for each objective (Simonsson et al. 2007), shown in Table 2.2.

Table 2.2 is commonly used to state the best case (5), worst case (1), something in between (3), and then filling in intermediate maturity levels, as available. A maturity assessment often provides context, in the way of an essay, for additional reasoning about the assessment level. For example, some approaches include filling out an objective assessment, or narrative, for what each of the 1–5 states look like, in order to get an initial assessment. Other approaches (i.e. NIST CSF) use the table to do a detailed system evaluation.

Modeling and simulation becomes a useful tool once the system security is assessed, providing a baseline, around the five steps of defensive cyber operations (Table 2.4). In addition, Table 2.3 overlaps with the Critical Security

Table 2.2 Maturity levels for policy implementation and process assessment.

Maturity level	Meaning
1	Initial/Ad hoc
2	Repeatable, but intuitive
3	Defined
4	Managed
5	Optimized

Table 2.3 Resilience lines of effort (LOEs).

Resilience step/level	5	4	3	2	1
Identify	The draft NIST CSF provides an example of this approach in evaluating the maturity levels for each of the respective resilience steps.				
Prevent					
Detect					
Respond					
Recover					

Table 2.4 Resilience steps and system security assessment.

Defensive cyber operation	System security assessment
Identify	• Inventory network for key assets • Prioritize network assets based on security evaluation
Prevent	• Percentage of controls in place • Cybersecurity policies in place • Cyber maturity model assessment level
Detect	• Internet facing protections in place • Internal security process in place • Log evaluation process for detecting anomalous events
Respond	• Readiness level of defensive cyber operation trained team (e.g. similar to Disaster Recovery/Continuity of Operations)
Reconstitute	• Backups available to install/resume system operation

Controls (SANS Institute 2006) are used to outline the identification and preventive steps necessary to do the initial network identification and protection. Examples are provided in Table 2.4.

As shown in Table 2.4, simple yes/no questions, or percentage of satisfaction for standard questions about precautions taken to secure a cyber system, provide the initial "scenarios," or use cases, to evaluate systems for secure cyber operation. Varying the use cases, to capture the scope of the issue at hand, however, can rapidly become more complicated. Ideally, use cases provide "outer bounds" for the respective system conditions, giving the evaluator an example of system performance at these respective boundaries.

2.2.2 Use Cases/Scenarios

Frameworks, designed for assessing actual systems, give the modeler an understanding of the current system structure, and provides context to describe system behavior. Use cases and/or scenarios are used to show strengths, weaknesses, and performance scaling of the system of interest over its scope of implementation. These use cases could be chosen from any of the people, policy, process, or technology system domains, or their respective combinations, where there is a perceived weakness to be tested against. In addition, specific threat behavior might be modeled to determine our system's attack susceptibility.

Using modeling and simulation to develop real-world cyber scenarios, scenarios that incorporate both the scope and scale of networks "in the wild," includes technologies and configurations that can easily span multiple generations of information technology. It is a challenge to measure risk in these complex systems, where a clear and accurate inventory of the technologies in defended systems is absent. One approach is to inventory a system of interest and compare component-level system descriptions with well-known weaknesses and vulnerabilities (Table 2.5) in order to develop in-depth assessments.

Security analysis leveraging Table 2.5's databases is used by tools (e.g. CAULDRON [Jajodia et al. 2015]) to prioritize system updates, thereby increasing overall resilience.

2.3 Cyber Insurance Approaches

In 2015, the Lloyd's of London study, "Business Blackout," (Maynard and Beecroft 2015) showed a potential 93 million Americans, across 10 states and the District of Columbia, without power due to a cyberattack; costing an estimated $243 billion; $1 trillion in the most stressing scenario (Figure 2.3).

One reason to look at insurance analyses is because insurers necessarily deal with the "as is," real, world, where clients want to transfer their cyber risk through an insurance policy. Reading the appendices of "Business Blackout" (Maynard and Beecroft 2015), there was noticeable variety in

- Countries attacked
- Facilities attacked
- Types of attackers
- Attacker technologies
- Attack effects

In addition, "Business Blackout" provides a picture that almost any vulnerability is liable to be exploited – a wake-up call as to what is possible. Expanding on insurance modeling as a template for describing cyber risk, Bohme and Schwartz (2010) provide an excellent background of cyber insurance literature and define a unified model of cyber insurance (i.e. cyber risk description) that consists of five components:

- the networked environment
- demand side
- supply side
- information structure
- organizational environment

First, the network topology plays a key role in affecting both security dependencies and failure correlations. For example, independent computers versus a fully connected computing network, two topology extremes, will result in vastly different security assessment and implementation challenges. In addition, system impacts will look very different, should a successful attack occur.

Table 2.5 Standard security description references.

Name	Description	Link
Common Vulnerabilities and Exposures (CVE)	Contains the list of known information security vulnerabilities and exposures.	https://cve.mitre.org/
National Vulnerability Database (NVD)	Based on CVE dictionary, NVD is the basis for constructing of attack graphs via known vulnerabilities.	https://nvd.nist.gov/
Common Vulnerability Scoring System (CVSS)	Open and standardized vulnerability scoring system.	https://nvd.nist.gov/vuln-metrics/cvss
Common Weakness Enumeration (CWE)	Contains a unified, measurable set of software weaknesses. Usage of the database of weaknesses can improve the quality of the zero-day-based attack graph generator module.	https://cwe.mitre.org/
Common Platform Enumeration (CPE)	Provides a unified description language for information technology systems, platforms, and packages.	https://scap.nist.gov/specifications/cpe/
Common Configuration Enumeration (CCE)	Gives common identifiers to system configuration issues.	https://nvd.nist.gov/config/cce/index
Common Attack Pattern Enumeration and Classification (CAPEC)	Helps to capture and use the attacker's perspective. Usage of attack patterns allows applying sequences of known and zero-day vulnerabilities in one attack action.	https://capec.mitre.org/
Common Remediation Enumeration (CRE)	Defines a security-related set of actions that result in a change to a computer's configuration.	https://scap.nist.gov/specifications/cre/
Repository of Industrial Security Incidents (RISI)	RISI, a database of industrial controls anomalies, was developed to share data across the research community to prevent future cyber anomalies on operational technology	http://www.risidata.com/About

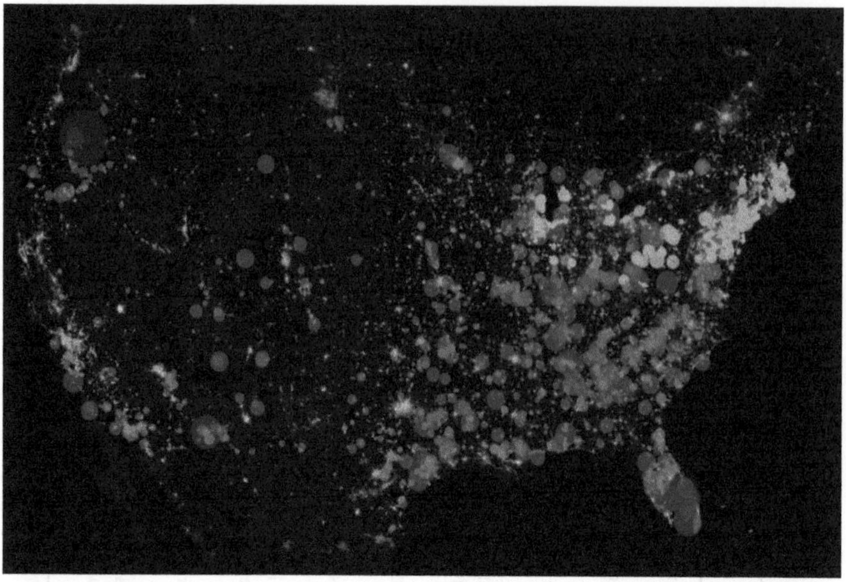

Figure 2.3 Business blackout due to US east coast grid compromise (Maynard and Beecroft 2015).

An additional insurance consideration includes using models of the defended network, potential impact, and defense cost estimates for both resilience and potential asset loss.[3] One challenge is for cyber insurance companies to price their offerings. Pricing can correlate to risk (Black and Scholes 1973), providing a view as to how well an insurance company believes it understands cyber risk. A recent study (Sasha Romanosky) looks at how insurers do business by evaluating several policies to see how companies evaluate risk. In regard to pricing, or rate schedules, there is a surprising variation in the sophistication of the equations and metrics used to price premiums. Many of the policies examined used a very simple, flat rate pricing (based simply on expected loss), while other policies use more parameters, such as the firm's asset value (or firm revenue), or standard insurance metrics (e.g. limits, retention, and coinsurance), and industry type. More sophisticated policies include specific information security controls and practices as collected from the security questionnaires.

---

3 Insurance companies consider, among other issues, the competitive landscape of insurers, contract design (premiums, fines), and the company's own risk preferences. Additional environmental considerations include issues such as regulatory forces that may exist to mandate insurance, disclosure requirements in the event of a loss, and the effect of outsourced security services and hardware and software vendors on a firm's security posture.

The seemingly capricious pricing model of cyber insurance is partially due to the seemingly random demand signal that current cyberattacks provide. We are in a new era, with nefarious cyber actors ranging from States to anonymous hackers, where the loss process that an insurance company constructs can only really be based on the business process value at risk[4] (Sasha Romanosky).

2.3.1 An Introduction to Loss Estimate and Rate Evaluation for Cyber

In performing quantitative risk assessment, insurance companies often use some form of an a single security incident, with the annualized loss expectancy (ALE) to justify the cost of implementing countermeasures to protect an asset. This may be calculated by multiplying the single loss expectancy (SLE), which is the loss of value based on a single security incident, with the annualized rate of occurrence (ARO), which is an estimate of how often a threat would be successful in exploiting a vulnerability (Equation 2.1).

$$ALE = ARO \times SLE$$

Equation 2.1: Annualized loss expectancy

In addition to traditional insurance assessment approaches, the FAIR model is an international standard[5] for computing value at risk, and is a practical framework for understanding, measuring, and assessing information risk, enabling well-informed decision making. In addition, the FAIR model is supported by a tool (i.e. Risk Lens) and a community for quantifying and managing cyber risk. In addition to the FAIR model, Douglas Hubbard's "How to measure anything in cyber security risk" (Hubbard et al. 2016) provides background on cyber measurement approaches, along computer-based approaches,[6] for demystifying cyber issues.

2.4 Conclusions

Insurance modeling is used here to provide a baseline for the "as is" of cyber security assessment. While assessments will leverage standard frameworks (e.g. NIST CSF) and risk evaluation approaches (e.g. ISO 31000 based), their goal of estimating system cyber resilience is the same. Leveraging maturity

4 Romanosky (2016) recently did an evaluation of current cyber events, with a goal of evaluating framework utility from an insurer's point of view. The paper estimates that an average cyber incident costs $200 000, averaging 0.4% of a company's estimated annual revenues.

5 http://www.fairinstitute.org/an-international-standard

6 http://www.howtomeasureanything.com/cybersecurity/

models is used to determine defensive capability across the respective lines of effort, providing a comprehensive approach for understanding an organization's current cyber security posture.

2.5 Future Work

In addition to the assessment frameworks and tools reviewed here, DoDAF architecture frameworks are an additional tool for documenting systems for cyber security evaluation (Hamilton, 2013; Richards 2014). Leveraging system understanding, through DoDAF architectures, along with threat evaluation, was recently done in a US Air Force system assessment approach (Jabbour and Poisson 2016).

2.6 Questions

1. Should an organization approach cyber risk, and subsequent insurance, with a different strategy than it approaches standard property and casualty (P&C) and other insurable business risk?
2. How is cyber risk transferred, now?
3. How much of cyber risk is usually accepted (vice transferred) by most organizations now?
4. When would a great resilience process provide a bad assessment?

3

Introduction to Cyber Modeling and Simulation (M&S)

Emergency responders (e.g. police, fire fighters, etc.) commonly use training exercises to develop both individual and team skills for known scenarios. These exercises, simulations of real events, are often categorized as "modeling and simulation," with simulacra of real entities composing the "models" in these events. Cyber defenders' use of M&S is relatively new.

As described in Chapter 2, analytic models are used to evaluate cyber system risk via assessment frameworks. Combining these legacy IA frameworks with developing cyber modeling theory provides a foundation for tools that perform the "what if" analyses enabling a science of cyber security.

3.1 One Approach to the Science of Cyber Security

Cyber M&S will be the tools through which future engineers and technologists practice a Science of Cyber Security. Kott (2014), for example, provides a cyber description based on a defense against malicious software with the following definition:

> "... the domain of science of cyber security is comprised of phenomena that involve malicious software (as well as legitimate software and protocols used maliciously) used to compel a computing device or a network of computing devices to perform actions desired by the perpetrator of malicious software (the attacker) and generally contrary to the intent (the policy) of the legitimate owner or operator (the defender) of the computing device(s)."

An Introduction to Cyber Modeling and Simulation, First Edition. Jerry M. Couretas.

In addition, Kott (2014) notes that the key objects of research in cyber security should be:

- Attacker, A, along with the attacker's tools (especially malware) and techniques T_a.
- Defender, D, along with the defender's defensive tools and techniques T_d, and operational assets, networks, and systems N_d.
- Policy, P, a set of defender's assertions or requirements about what event should and should not happen; simplifying to the cyber incidents, I, that should not happen.

Kott generalizes cyber security to Equation (3.1)'s 4-tuple, M, as shorthand for expressing what we might expect to encounter in a cyber incident:

$$M = \left\{ I, T_d, N_d, T_a \right\}$$

Equation 3.1: Network model taxonomy description.

I: cyber incidents, events that should not happen
T_d: defender's defensive tools and techniques
N_d: defender's operational assets, networks, and systems
T_a: attacker's tools (e.g. malware) and techniques

Equation (3.1) provides an extensible representation for an overall cyber modeling framework, accounting for a behavioral view of cyber security, at a higher level of abstraction than the current Confidentiality, Integrity, and Availability (CIA) models of network defense. The value of Equation (3.1) is that an analyst can see the entire cyber problem space without getting lost in details, a common challenge with constructive modeling.

Kott's 4-tuple compliments the recent National Academy of Sciences study (Millett et al. 2017) findings in looking at institutional improvements required to develop a science of cyber security. Key findings include:

- **Enabling Research** – current, high-frequency, publishing rhythm can leave current literature conclusions a challenge to duplicate; the suggestion is to do longer-term projects that provide results that enable the readers to replicate results.
- **Cyber as an Interdisciplinary Field** – social science aspects, in dealing with the human interface to computer-based systems, affect all aspects of cyber security, from policy through technology; one suggestion is that new doctoral students should have a double major that includes both a technical and social science discipline, if working on cyber security.

In addition to Kott's concise 4-tuple, and the National Academies' recommendations on cyber, Couretas (2017) provides an overview of M&S maturity for the developing science of cyber security in.

3.2 Cyber Mission System Development Framework

Kott's 4-tuple, outlining the space for modeling a science of cyber security, is complemented by a conceptual model that adds mission context for contemporary cyber operations. For example, cyber mission systems, elements covered in the DoD's Cyber Science and Technology (S&T) Priority Steering Council Research Roadmap, are shown in Figure 3.1, and span from the effects desired (right-side of Figure 3.1) to the sensors and situational awareness (left side of Figure 3.1). In addition, desired architectural characteristics (e.g. trust, assuredness, and agility) are described in a hierarchical fashion as the system builds through the center of the diagram.

Figure 3.1's cyber mission system components provide the high-level elements and capabilities desired in an overall system. In addition, Figure 3.1 is a conceptual model, laying out the effects desired from a constructed system. The middle tiers provide example metrics that the system will be designed to accomplish. The left side is monitored via an experimental frame during development, through real-world sensors in practice.

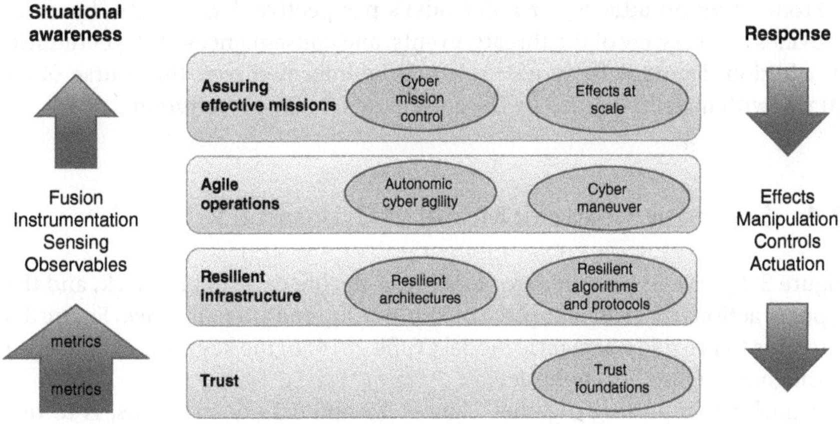

Figure 3.1 DoD's cyber S&T priority steering council research roadmap (King 2011).

3.3 Cyber Risk Bow-Tie: Likelihood to Consequence Model

One way to look at Figure 3.1 is as an overall architecture description, each instance of which will require a system security evaluation similar to Figure 3.2's "bow-tie," which shows how different controls and countermeasures fit along hypothetical attack paths (Nunes-Vaz et al. 2011, 2014).

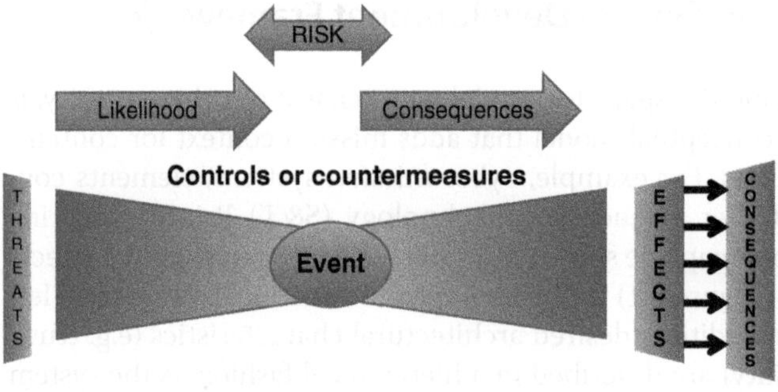

Figure 3.2 Cyber risk "bow-tie" – prevention, attack, and remediation.

As shown, the left side of Figure 3.2 works to minimize the risk of a cyberattack, leveraging (Chapter 2, Table 2.1)'s system risk characterization, while the right side of Figure 3.2 provides the resilience, or consequence management, required to handle a cyberattack currently under way. In addition, Figure 3.2 was developed with the ISO 31000 risk standard in mind.

From either an attacker or a defender's perspective, Figure 3.2's "bow-tie" provides an overview of the threats, events, and consequences of a cyberattack. In addition, Figure 3.2's attack cycle will be informed over the course of an attack, with metrics defined by the enterprise's policy prescription.

3.4 Semantic Network Model of Cyberattack

Figure 3.2 provides a method for looking at the life cycle of an attack, and the types of actions that will take place before, during, and after an attack. Figure 3.3 attempts to provide a semantic model (Yufik 2014) of the key entities leveraged throughout Figure 3.2's bow-tie.

A goal, when putting together Figures 3.2 and 3.3's descriptions, is to test each of the respective controls (Figure 3.2's left side), or attack counter measures (Figure 3.2's right side), as a means of Course of Action (COA) evaluation. This kind of testing, currently performed on real equipment, or emulators (e.g. a cyber range), is a key area where modeling may contribute to COA strategy evaluation (e.g. automated defenses, moving target representations) (Okhravi et al. 2013a, b). Leveraging the overall flow of Figure 3.2, we will use Figure 3.4 for cyber model construction efforts.

As shown in Figure 3.4, our approach begins with scenarios, looking at courses of action (COAs) and associated models that may apply. Scenarios, as provided in Figure 3.4, are proposed here as a more generalized structure than the use cases (Figure 3.5) that ideally guide the way for the categorizing and measuring cyber phenomena.

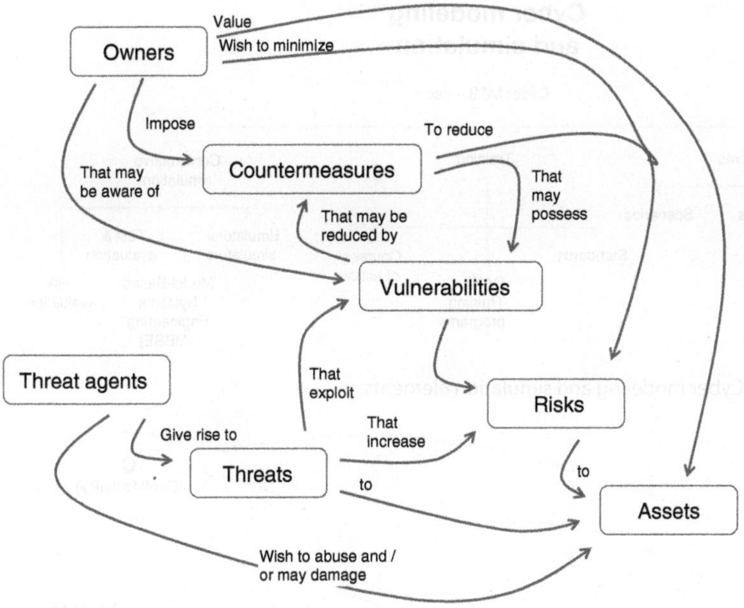

Figure 3.3 Semantic network of current and anticipated threats (Yufik 2014).

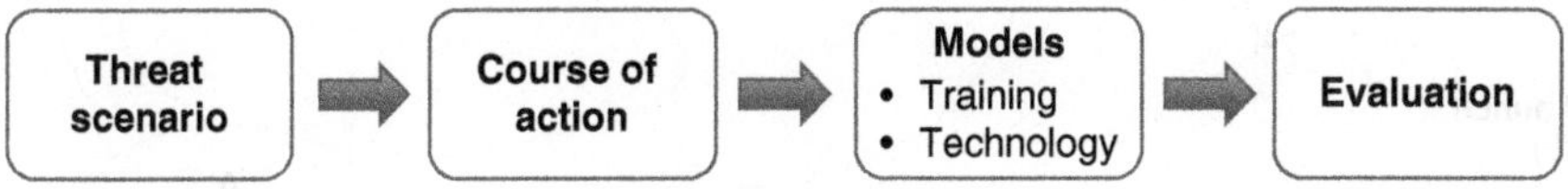

Figure 3.4 Scenarios through model development approach.

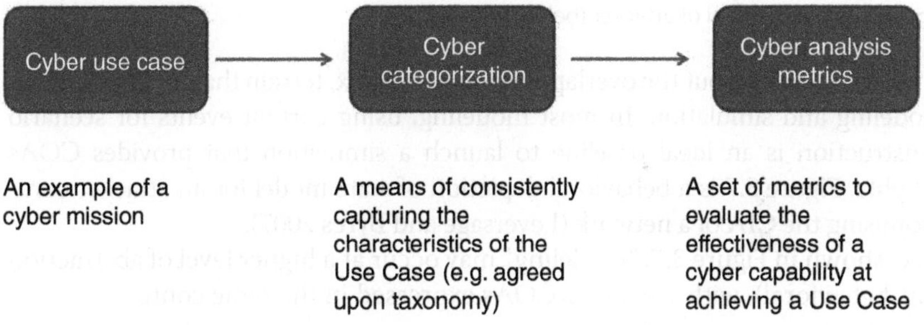

Figure 3.5 Cyber analysis elements.

While Figure 3.5 provides an idealized distillation of capturing cyber phenomena, an overall diagram that includes each of the cyber M&S elements is shown in Figure 3.6.

Cyber categorization might be done by something like MITRE's Att@ck framework (https://attack.mitre.org/wiki/Main_Page).

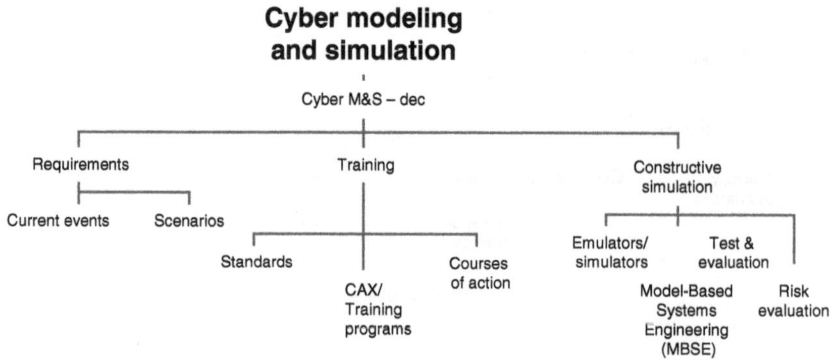

Figure 3.6 Cyber modeling and simulation elements.

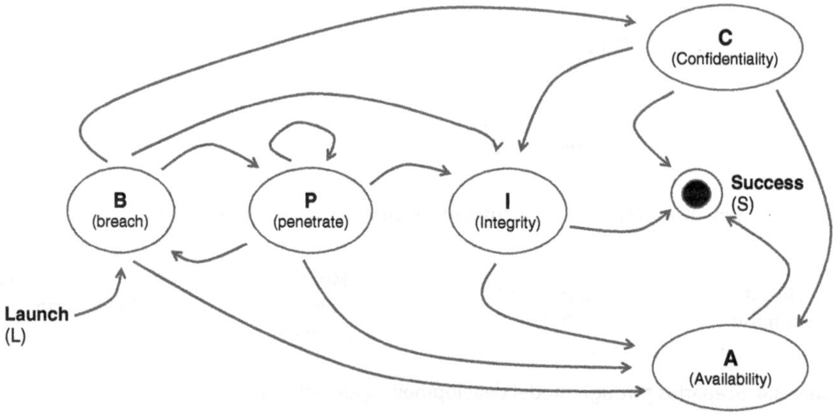

Figure 3.7 State model of attacker (behavioral example).

Figure 3.6 brings out the overlapping, and complex, terrain that makes up cyber modeling and simulation. In most modeling, using current events for scenario construction is an ideal baseline to launch a simulation that provides COAs insights. Figure 3.7 is a behavioral depiction of state model for an attacker compromising the CIA of a network (Leversage and Byres 2007).

As shown in Figure 3.7, "modeling" may occur at a higher level of abstraction (e.g. behavioral), with scenarios/COAs expressed in the same context.

3.5 Taxonomy of Cyber M&S

A recent NATO taxonomy and literature review (Lange et al. 2017) for common types of models in cyber defense is shown in Table 3.1.

Table 3.1's model summary provides an overview of the types of cyber M&S applications observed in the "Model-Driven Paradigms for Integrated

Table 3.1 Taxonomy and models for cyber defense.

Modeling type	Description
Emulation	Emulation (often with simulation) of networks: actual hardware, software, and humans (e.g. cyber ranges)
Training	Training-focused simulations: presenting to human trainees the effects of a cyberattack without modeling underlying processes
M&S of human cognitive processing of cyber events and situations	Perception, recognition, situational awareness (SA), and decision making
M&S of attack progress and malware propagation	• Attack–graph-based approaches • Epidemiology analogy (e.g. Susceptible, Infected, Recovered [SIR])
Abstract wargaming	Game-theoretic model of cyber conflict without modeling the underlying processes of cyberattack and defense
Business process models	Defense, offense, and business processes, along with business information technology architecture, simulated for observing effects
Statistical models of cyber events	Cyber processes represented as, for example, equations of stochastic processes, and coefficients learned from real events, or a training data set
Two classes of models that support cyber modeling, but do not model cyber aspects	• Physical systems models to support modeling of cyber–physical effects • Network simulation models

Approaches to Cyber Defense" (NATO IST-ET-094) study Lange et al. 2017). One implementation of a cyber model, a first step, is constructive modeling of a cyber system for situational awareness.

3.6 Cyber Security as a Linear System – Model Example

At a slightly lower level of abstraction, a cyber model is developed through leveraging dependencies. This includes modeling incomplete and noisy observations via integrating Bayesian network, Markov, and state space models (Cam 2015). Cam's approach accounts for the inherent ambiguity in cyber environments and uses defined asset dependency and criticality to construct alternative mission paths. This includes leveraging observability to characterize the system state for assessing potential weaknesses and vulnerabilities; and proving the controllability to steer a network with some compromised components towards a desired state within finite time. For example, consider a network of N nodes/clients, where

$$N = G(t) + V(t) + C(t) + E(t) + F(t)$$

Equation 3.2: Cyber as a linear system

$G(t)$: the number of those nodes that do not have any known vulnerability at time t.
$V(t)$: the number of those nodes that have some known vulnerabilities at time t, but are not exploited yet.
$C(t)$: the number of those nodes that are compromised partially/fully through the exploitation of their vulnerabilities.
$E(t)$: the number of those nodes that are evicted due to their not being recoverable.
$F(t)$: the number of those nodes that have failed and do not operate due to physical failures.

We can control the states and operation of nodes by $P(t)$ and $R(t)$; we can measure $C(t)$ and $V(t)$.

input 1: $P(t)$; input 2: $R(t)$.
$R(t)$: recovery support services rate.
$P(t)$: patching support services rate.
$o(t_0)$: vulnerability occurrence rate.
$p(t_0)$: vulnerability patching rate.
$e(t_0)$: vulnerability exploitability rate.
$r(t_0)$: compromised systems' recovery rate.
$d(t_0)$: cyber-compromised node eviction rate.
$f(t_0)$: physical failure rate.

Figure 3.8, which could be looked at as an epidemiology model, provides a high-level view of system performance, with the potential for measuring both performance and effects based on current network state.

3.7 Conclusions

Dr. Cam's constructive modeling approach is one example of a roll-up description that provides for both system evaluation and situational awareness to ensure that system behavior is in-line with expected performance. This example fits nicely with the span of developing cyber models, from training through analytic failure analysis. These models are a valuable step forward in the construction of a Science of Cyber Security, as proposed by both the National Academy of Sciences and Dr. Alexander Kott.

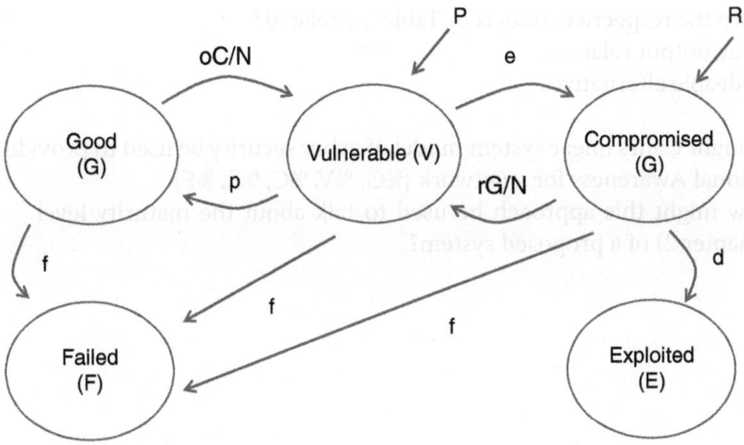

Figure 3.8 Cyber security as a linear system (Cam).

While the expanding scope of cyber modeling requires ongoing literature reviews to understand how the field is developing, significant progress has been made in recent years, as described by Dr. Kott's n-tuple model, more explicitly in Cam's linear system description. These models, along with the broader understanding of the attack lifecycle provided by the risk bow-tie, provide fertile terrain for the continuing use of M&S to leverage scenarios in testing, and evaluating proposed and operational systems.

3.8 Questions

1 Why is risk evaluation, as used in Information Assurance, not part of the standard domain of M&S?

2 Name two examples of resilient architectures and/or resilient algorithms and protocols that cyber M&S can help evaluate for effectiveness or performance?

3 How is cyber mission control achieved now? Situational Awareness?

4 How might the cyber risk bow-tie (Figure 3.2) be modeled?
A Analytically
B With event-based modeling
C With knowledge elicitation techniques

5 In using the "state Model of an Attacker," (Figure 3.7) is it true that "Success" is reachable from each of the C, I, or A phases of an attack?

6 How are the respective models in Table 3.1 related?
   A Input/output relations
   B Tradeable alternatives

7 How might Cam's linear system model of cyber security be used to provide Situational Awareness for a network {%G, %V, %C, %E, %F}
   A How might this approach be used to talk about the maturity level (Chapter 2) of a proposed system?

4

Technical and Operational Scenarios

As discussed in Chapters 2 and 3, describing "cyber" is a challenge, from evaluating business impact to modeling the technical underpinnings that compose the backbone of our critical systems. For example, Mandiant (FireEye 2017) recently reported malware detection and mitigation numbers:

> "Fortunately, we're seeing that organizations are becoming better at identifying breaches. The global median time from compromise to discovery has dropped significantly from 146 days in 2015 to 99 days in 2016, but it is still not good enough. As we noted in M-Trends 2016, a Mandiant Red Team can obtain access to domain administrator credentials within roughly three days of gaining initial access to an environment, so 99 days is still 96 days too long."

Developing technical and operational scenarios is an activity that spans from policy to technical implementation in determining the controls and indicators used in proper system evaluation. A popular approach for performing high-level evaluation includes threat modeling, which provides opportunities for future scenario and Course of Action (COA) use cases. For example, PASTA™ (Velez and Morana 2015) (Table 4.1) provides an overall methodology for threat evaluation that could serve M&S as an overall approach.

Table 4.1's PASTA, with similarities to the NIST SP 800 approach (Chapter 2, Table 2.1), is an example of a high-level analysis approach for developing future, baseline, scenarios, and subsequent Courses of Action (COAs). Leveraging end-to-end processes for system evaluation will be aided by system decompositions for follow-on evaluation. The ARMOUR Framework (DRDC (Canada) 2013a, b) provides example technical and operational scenarios with the aim of supporting an overall cyber framework.

An Introduction to Cyber Modeling and Simulation, First Edition. Jerry M. Couretas.

Table 4.1 Stages of Process for Attack Simulation and Threat Analysis (PASTA) threat modeling methodology.

Define objective	• Identify business objectives • Identify security and compliance requirements • Technical/business impact analysis
Define technical scope	• Define assets • Understand scope of required technologies • Dependencies: Network/software (COTS)/service • Third-party infrastructures (Cloud, SaaS, Application Service Provider [ASP] Models)
Application decomposition	• Use cases/Abuse (misuse) cases/Define app entry points • Actors/Assets/Services/Roles/Data sources • Data Flow Diagramming (DFDs)/Trust boundaries
Threat analyses	• Probabilistic attack scenarios • Regression analysis on security events • Threat intelligence correlation and analytics
Vulnerability and weakness mapping	• Vulnerability database or library management (CVE) • Identifying vulnerability and abuse case tree nodes • Design flaws and weaknesses (CWE) • Scoring (CVSS/CWSS)/Likelihood of exploitation analytics
Attack modeling	• Attack Tree Development/Attack Library Management • Attack node mapping to vulnerability nodes • Exploit to vulnerability match making
Risk and impact analysis	• Qualify and quantify business impact • Residual risk analysis • Identify risk mitigation strategies/develop countermeasures

4.1 Scenario Development

Protective cyber scenarios take a variety of forms. For example, Table 4.2 provides a set of potentially life-threatening examples where cyber operators likely had minimal notice to determine a real-time COA.

As shown in Table 4.2, cyber scenarios occur over a period of time, usually in stages, including both technical and operational elements in the detect–mitigate–recover phases of a resilience scenario. To assist in technical evaluation and operator training, Canada's ARMOUR (DRDC (Canada) 2014a, b) cyber technical demonstrator (TD) developed technology-specific approaches, including "proactive" and "reactive" scenarios, as shown in its concept of operations (CONOPS) (Table 4.3).

Table 4.2 Operational examples.

Date	Scenario example
April 2016 until at least February 2017	Operation Electric Powder (ClearSky Research Team 2017) 1) Spear phish (PC, Android phone) 2) Directed to watering hole (Facebook) Attempt to penetrate Israel Electric Company (IEC)
2014	German Steel Mill Cyber Attack (Lee et al. 2014) with confirmed physical damage
February 2013 to April 2014	Dragonfly (Symantec 2014): Cyber espionage attacks Against energy suppliers • 2/2013–6/2013 Spam Campaign • 9/2013–Lightscout exploit kit used • 5/2013–4/2014 Watering Hole attack A newer approach used by the attackers involves compromising the update site for several industrial control system (ICS) software producers. They then bundle Backdoor.Oldrea with a legitimate update of the affected software. To date, three ICS software producers are known to have been compromised. The Dragonfly attackers used hacked websites to host command-and-control (C&C) software.
August to September, 2013	Rye Dam (New York) (NEWSWEEK 2016) – Threat actors accessed the Supervisory Control and Data Acquisition system, which connects to the Internet through a cellular modem – after allegedly obtaining water-level and temperature information, could have operated the floodgate remotely if it had been operating at the time.

As shown in Table 4.3, proactive scenarios are used to evaluate how the network responds to anomalies, or time to detect (T_{detect}) anomalous devices and configuration changes. These are sometimes called technical scenarios; similar to what is evaluated via critical security controls (CSCs). Reactive scenarios, on the other hand, are usually called operational simulations, often training-focused, and are used to perform standard Disaster Recovery/Continuity of Operations (DR/COOP) enterprise evaluations.

Scenarios, therefore, are inherently context-dependent, in that applying CSCs should protect a system from obvious threats, with training to maintain both awareness and responsiveness should an attacker gain access.

4.1.1 Technical Scenarios and Critical Security Controls (CSCs)

Technical scenarios primarily deal with network anomalies. As a preventive example, the Australian Signal Directorate's "Top 4" is popular for its reported ability to prevent 85% of cyberattacks (Defense). Similar to the ASD opposition force is the NIST 800-53-based CSCs.

Table 4.3 Proactive and reactive ARMOUR scenarios (DRDC (Canada) 2014a, b).

ARMOUR scenario	Description
Proactive	1) Addition of new hosts 2) Addition of new network device (switch, router, etc.) 3) Addition of new security device (firewall, gateway, etc.) 4) Modification to existing network device 5) Modification to existing security device
Reactive	Once an asset has been identified with an exploited vulnerability, ARMOUR provides the operator with the capability to identify potential attack paths or attack vectors to other assets that may have been exposed. This attack path can provide insight into other similarly affected hosts and can also indicate where this exploit, or a related exploit, could be used to gain access to another network connected host in the topology. With this ability to uncover the potential attack vectors, ARMOUR provides the operator with a complete understanding of the potential capabilities that the observed exploit could provide to the attacker. Once the attack graph is generated, COAs are provided to the operator to resolve the vulnerabilities thereby mitigating the propagation of the attack any further. Simulation of the COAs demonstrates to the operator the impact of implementing the risk mitigation. The COAs implemented could include removing the vulnerability from the attack point (initially infected asset) and/or removing the vulnerability from assets further down in the attack path.

A clear advantage to using operational scenarios, as shown in Table 4.3, is that the evaluations are in numbers (number of nefarious nodes, time to respond, etc.). Operational scenarios are more challenging with the story line needing to be developed for the specific threat of interest.

4.1.2 ARMOUR Operational Scenarios (Canada)

ARMOUR (DRDC (Canada) 2013a, b, 2014a, b) was a Canadian effort to develop an architecture-based framework, leveraging cyber models, to build a test bed for training and technology evaluations. The ARMOUR framework performs operational scenarios using the same underlying technical architecture, changing the focus from measurable network goals to more human-oriented evaluations (Table 4.4).

As shown in Table 4.4, ARMOUR scenarios are practical, mirroring actual network events that are easily duplicated on real or emulated networks. In addition, operational scenarios often involve a story line, usually from a real incident (Table 4.2), that is generalized to a training objective (Kick 2014). As shown in Figure 4.1, cyber events are developed with both the event goals and the estimated environment in mind.

Table 4.4 ARMOUR operational scenarios (DRDC (Canada) 2014a, b).

Scenario	Description
User Identification and Authentication	The User Identification and Authentication operation represents the system interactions between services/modules during an operator login to the data presentation framework.
User Data Request	The User Data Request Operational Scenario describes the system and service interactions and data flows for the situation where a user opens a presentation view and makes a request to view stored data.
Network Data Collection and Presentation	The Network Data Collection and Presentation scenario depicts an information flow for the collection, normalization, validation, storage, and presentation of network information from a data source. The contextual operation represents a generic flow of data and can be applied to virtually any data source (each data source will have at least one individual data source connector).
Reaction to Events	In this scenario, a link between a host and router/switch will become saturated and will require the user's intervention. The user will select the intervention widget and be able to "drag n drop" an alert from the Alerts component to the intervention component. The intervention component will be populated with relevant remediation methods based on the alert type being displayed. Upon selection of an intervention method, the system will generate a rule request to fix the issue.

Each of the Scenarios in Table 4.4 will require a structured process to determine how well the team did in defending their system. Figure 4.1 provides a standardized approach for conducting cyber events for security evaluation.

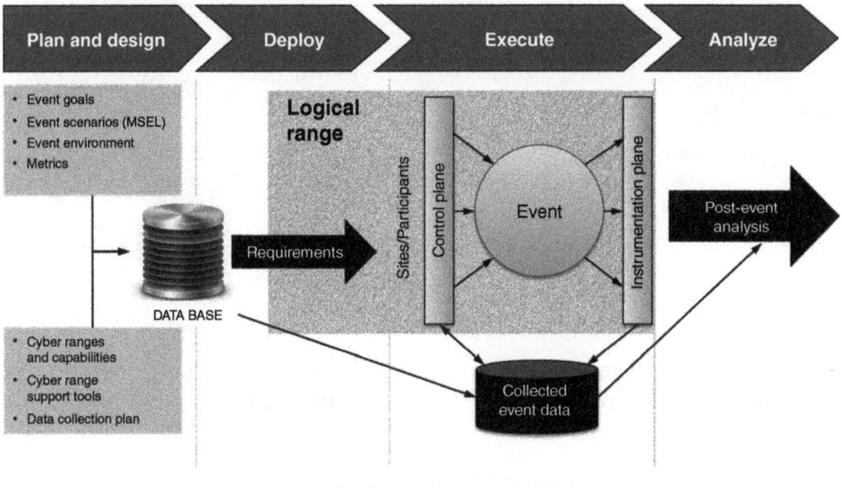

Figure 4.1 Cyber-range event process overview.

Figure 4.1 (Damodaran and Couretas 2015) provides an example exercise flow, often used with a scenario designed to emulate real-world events; e.g. operational examples for ICSs shown in Table 4.2.

4.2 Cyber System Description for M&S

Risk evaluations provide a system overview, and potential baseline, for a system's estimated susceptibility to known cyber issues. Generalizing on these assessments, usually static evaluations, in developing a repeatable and valid cyber description for M&S, is a challenge. Many descriptions attempt to bridge current cyber's Information Assurance (IA) foundations to provide approaches that span from IA to M&S.

4.2.1 State Diagram Models/Scenarios of Cyberattacks

One approach (Leversage and Byres 2007) is to (i) decompose the network into its respective sections and (ii) use CIA language to describe the course of an attack (Figure 4.2).

Figure 4.2's method provides an approach for answering strategic questions. For example, the likelihood of succeeding along one of the CIA paths, the time it takes, or any associated operational costs to improve threat path defense.

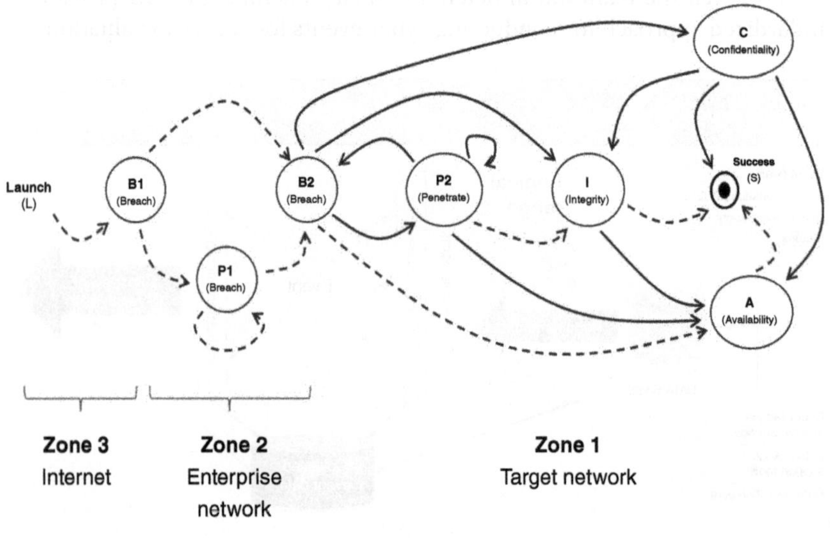

Figure 4.2 Attack path model using "CIA" system states.

4.2.2 McCumber Model

While Figure 4.2 provides a path model for a cyberattack, it is a natural next step to ask for more detail concerning the underlying system and its security posture. The McCumber model, well known to information security researchers, is reviewed here for cyber M&S scenario development. In addition to describing cyber security processes, the McCumber model "protects" the confidentiality, integrity, and availability (CIA) of mission systems "during" data storage/processing/transmission, while "using" technology/people/procedure and policy (Figure 4.3). From an M&S perspective, the McCumber model provides a conceptual approach to explore the impact of cyber activity on technology (i.e. a physical system) as well as people (i.e. behavior). In modeling a particular cyber phenomenon, the model captures all the parameters that must be addressed within the M&S environment. For a technology, the effect must be adequately modeled to represent its storage and processing capability during a transmission as well as all activities taken to protect the data from cyber activities.

The functional aspects of the McCumber model dovetail with the more structured requirements of M&S, generally, and scenario development, more specifically. An extension to the McCumber model includes:

- **Authentication:** guaranty vis-à-vis the destination that the information's origin/content is confirmed and certified as such. Each party to an exchange of information on both sides should be able to guarantee the identity of the other parties involved.
- **Non-repudiation:** guaranty vis-à-vis the origin, that the information reached the destination intact and unaltered. It is a guaranty that the information has been delivered to the destination, preventing the recipient from later denying receiving it. Non-repudiation protects against counterfeit information.

Figure 4.3 McCumber model.

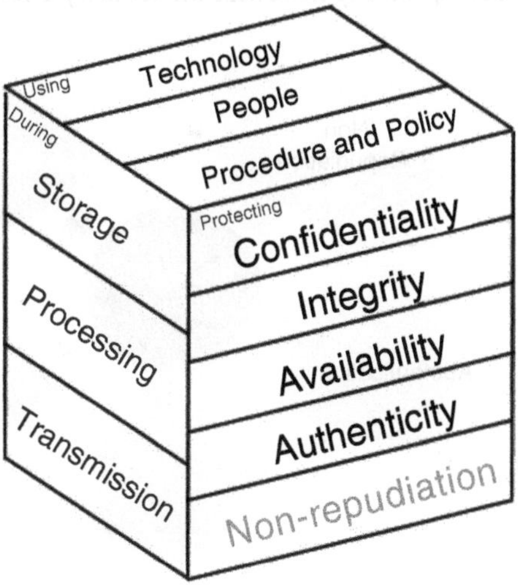

In addition to providing data provenance, the McCumber Cube provides a straightforward approach for looking at data, at rest or in transmission, to add a layer of technical detail to the IA CIA evaluation (Figure 4.2). Modeling both forms of data is of interest for scenario development.

4.2.3 Military Activity and Cyber Effects (MACE) Taxonomy

In addition to the McCumber model's more detailed description of the cyber terrain, we further narrow the scope of our cyber scenario development efforts with the Military Activity and Cyber Effects (MACE) taxonomy (Bernier 2015), which consists of six main categories:

- **Attack Types:** covers the most significant types of cyber-attacks.
- **Levels of Access:** describes the different levels of access to the targeted system or network required to launch a type of attack.
- **Attack Vectors:** includes the methods and tools used to infiltrate computers and install malicious software.
- **Adversary Types:** identifies the various types of cyber attackers.
- **Cyber Effects:** describes the effects that can be produced in the cyber environment by employing the various cyber-attacks.
- **Military Activities:** includes the military effects that can be produced in the cyber environment.

In addition, the MACE taxonomy provides a means for cross-referencing cyber effects with military activities to provide an overall impact estimate:

$$\text{Impact} = (\text{Military activity}) \times (\text{Cyber effect})$$

Leveraging MACE, we develop attack types, with the goal of looking at their corresponding information security effects (Figure 4.4).

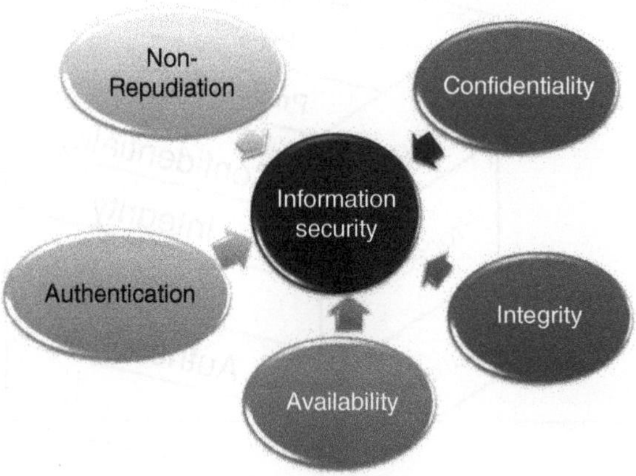

Figure 4.4 Components of information security.

Table 4.5 Cyber effects and military activities.

Cyber effects and military activities	Description
Cyber effects	• Interruption (Availability) • Modification (Integrity, Authenticity) • Degradation (Availability) • Fabrication (Integrity, non-repudiation) • Interception (Confidentiality) • Unauthorized use (not considered)
Military activities	• Deny, Degrade, Disrupt, Destroy, Digital Espionage • Defensive cyber operations • Offensive cyber operations o Cyberattack (Deny, Degrade, Disrupt, Destroy, Digital Espionage) o Cyber exploitation (Access, Gather Data, Digital Espionage)

From the M&S perspective, these six categories should be considered. In particular, the cyber effect combined with the military activity represents the impact a particular cyber threat may have on an operation. Table 4.5 captures the relationship between various cyber effects and military activities for consideration during scenario development.

While MACE provides an initial approach for providing a cyber/military effects description, the Cyber Operational Architecture Training System (COATS) leveraged actual range effects to inform cyber training simulation.

4.2.4 Cyber Operational Architecture Training System (COATS) Scenarios

COATS explores methods for using M&S to support training. The COATS program demonstrated several interoperability approaches for supporting M&S to include the exploring of extensions to data models to specifically model cyber effects. The objects of the COATS program are to:

- Enable synchronous execution of traditional training and cyber operations.
- Accurately model and simulate traditional training and cyber events/interactions.
- Draft interoperability guidelines for cyber-traditional federation.
- Distribute realistic cyber effects to the entire staff.

4.2.4.1 Cyber M&S Operational View Architecture (OV-1) (COATS Example)

For all the scenarios described, the operational architecture remains the same; the cyber range provides a safe environment to deploy a cyber operation. The key parameters representing the cyber operation are identified, captured, and represented in the model. The cyber effect is then transitioned from the cyber range, to a training environment, to emulate an actual cyberattack on an operator's workstation. This general approach has proven effective for training operators in identifying and responding to cyber activities. Figure 4.5 depicts the generalized OV-1.

As shown in Figure 4.5, a cyber range environment is used for mission operator training. We will next provide a few examples that leverage several scenarios from the COATS (Wells and Bryan 2015) project.

COATS was evaluated via Table 4.6's four scenarios, which presented both cyber and mission effects. Both the full motion video degradation and the command and control examples dealt with packet loss (i.e. Integrity in the CIA triad) in simulating performance deterioration, from a transmission and process standpoint, respectively. In addition, the SYN Flood (i.e. denial of service attack) and data diddling examples (i.e. critical asset blue screen of death) were both processing phase attacks, requiring more refined information (i.e. McCumber Cube description) on the part of the attacker.

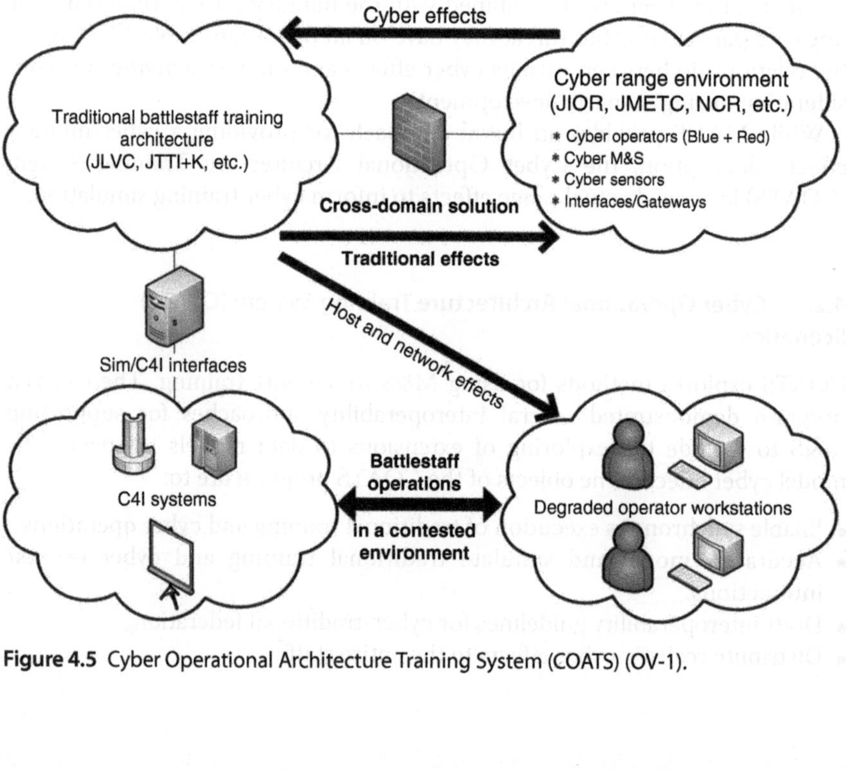

Figure 4.5 Cyber Operational Architecture Training System (COATS) (OV-1).

Table 4.6 Cyber Operational Architecture Training System (COATS) scenarios.

Scenario	Description
Computer Network Attack (CNA)	Live red CNA against virtual blue systems to demonstrate virtual host degradation effects on live operator workstations.
Node Attack	Constructive red kinetic attack on a constructive blue communications facility to demonstrate C2 disruption effects on live operator workstations.
Distributed Denial of Service	Live red CNA on virtual blue systems to demonstrate virtual full-motion video degradation effects on live operator workstations.
Threat Network Degradation	Live blue CNA on virtual red networks to demonstrate constructive system degradation on constructive red systems.

One of the key takeaways of Table 4.6's four scenarios is the applicability of the McCumber model to cyber M&S scenario development. The McCumber model provides clarity on what is being protected (e.g. CIA), when (transmission, storage, processing) and how (technology/people/procedure). Using this approach provides a clear language for how and why scenarios are constructed for cyber modeling and simulation, clarifying some of the uncertainty now found in applying cyber to standard training models.

Leveraging Table 4.5's definitions, one example is of mapping the COATS (Wells and Bryan 2015; Morse et al. 2014a, b) vignettes to Table 4.7's cyber effects; along with attack examples and McCumber Cube (Figure 4.3) descriptions of how the attack may occur.

The Attack Types represent the cyber effect modeled in each scenario and the two columns, "Using" and "During" detail the systems (technology), people, and timing of the effect in the scenario. Cyber effects in Table 4.7's first column are the four specific vignettes developed by the COATS program. While the MACE and McCumber approaches capture cyber effects and system operations in Table 4.7, accounting for CIA in standard IA terminology, constructive modeling will likely occur at a lower level of description.

4.3 Modeling and Simulation Hierarchy – Strategic Decision Making and Procurement Risk Evaluation

As introduced by the COATS figure (Figure 4.5), understanding the combined technical (e.g. network anomaly) to mission effect is one of the primary goals of cyber M&S thus far. Rowe et al. (2017) provides a depiction of how M&S might support strategic decision making in Figure 4.6.

Table 4.7 Cyber effects and attack type examples.

Military activity	Cyber effect	Attack type	System performance effect	Using	During
Deny (degrade, disrupt, destory)	Interruption (Availability)	Full Motion Video (FMV) degradation	Latency, jitter, packet loss	Technology	Transmission
	Degradation (Availability)	Interrupt supply chain and/or force flow	ICS, Location fidelity	All	Transmission, Storage
	Interruption (Availability)	System Shutdown	Memory Utilization	Technology P&P	Processing
Manipulate	Modification (Integrity, Authenticity	Reduce Situational Awareness, interrupt/ delay C2	Packet loss	Technology	Transmission

While the COATS diagram (Figure 4.5) provides the mechanics for incorporating cyber effects into training simulations, Figure 4.6 works to clarify the taxonomy of events, including cyber, that help with developing decision points in both cyber and campaign models. In addition, Figure 4.6 leverages the risk bow-tie (Figure 4.7) when considering preventive and remediation control applications.

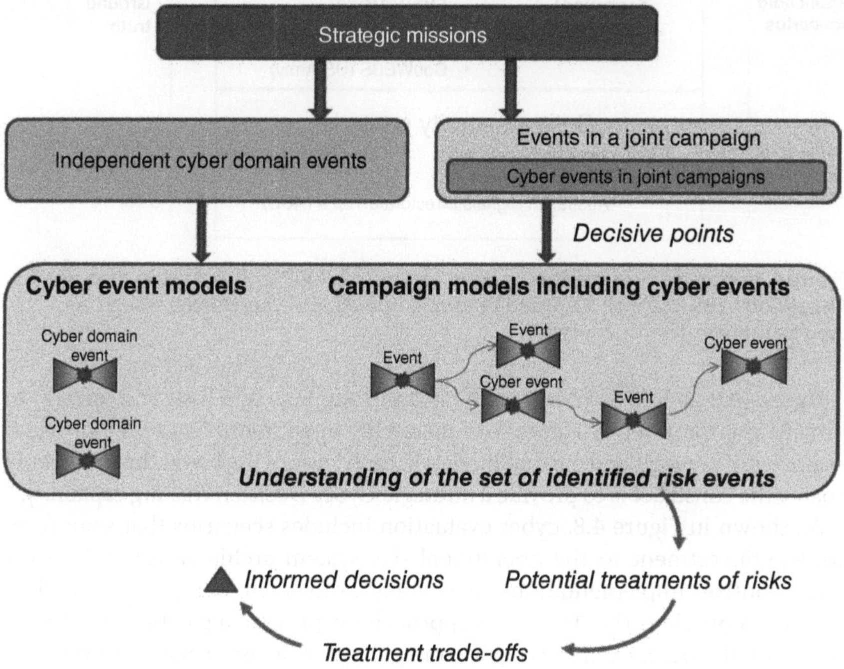

Figure 4.6 Cyber effects and mission evaluation Rowe et al. (2017) – http://journals.sagepub.com/doi/abs/10.1177/1548512917707077?journalCode=dmsa

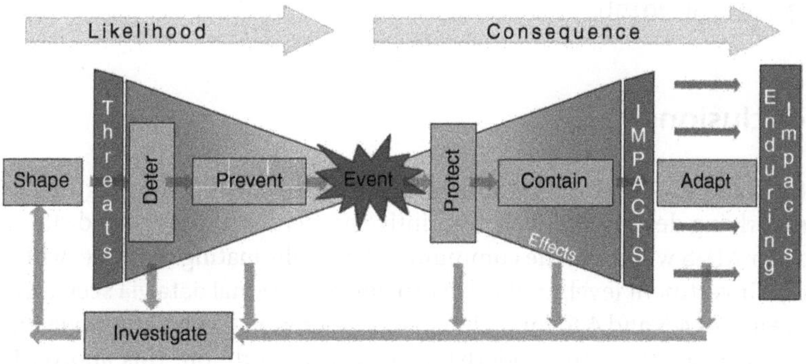

Figure 4.7 Risk bow-tie (Nunes-Vaz et al. 2011, 2014).

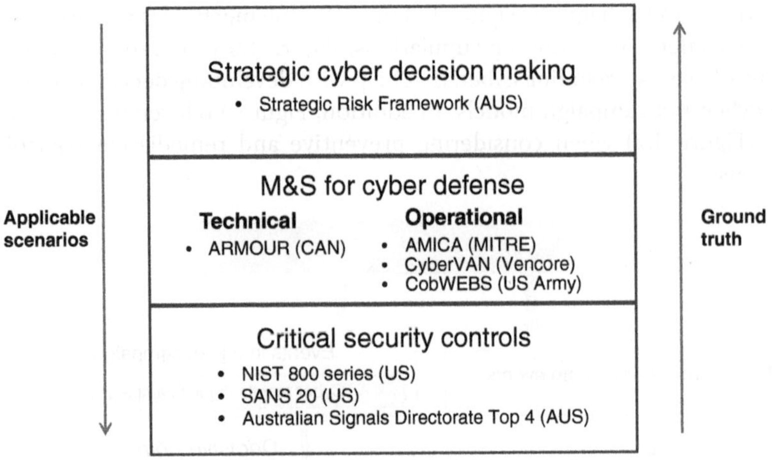

Figure 4.8 Strategic cyber decision making – leveraging M&S tools and cyber controls. US Army's CobWEBS (Marshall 2015) and Vencore Corporation's CyberVAN are models currently used to evaluate defense concepts.

Figures 4.6 and 4.7combine, in the form of the Strategic Risk Framework, to provide the top layer of Figure 4.8's hierarchy. In addition, Figure 4.8 provides examples of controls and current models at each layer of the hierarchy; the overall goal of the construct is to provide a Strategic Cyber Decision making capability.

As shown in Figure 4.8, cyber evaluation includes scenarios that span from strategy/investment to the operational (i.e. system architecture) and lower-level control implementation; leveraging both technology and training. Figure 4.8 provides the Australian approach for prioritizing cyber investment (Rowe et al. 2017), exemplifying a strategic cyber decision-making overview, used here for investment evaluation, leveraging the standard CSCs used by IT professionals to secure the network. M&S for cyber defense describes the frameworks (e.g. Canada's ARMOUR) and operational models (e.g. MITRE's AMICA [Noel et al. 2015]).

4.4 Conclusions

Figure 4.8's example of performing strategic portfolio evaluation leveraging the correct underlying descriptive and prescriptive models is one of the end states for how cyber M&S will serve the community. This culminating example, while at a strategic investment level, could also provide operational data via scenarios (Table 4.2, Figures 4.5 and 4.6) for technical and operational evaluations via the estimated performance of the underlying system. In addition, this approach spans from preventive and reactive, through technical/operational modeling, to strategic risk evaluation for an enterprise-level cyber system.

4.5 Questions

1 Name some common approaches for describing Figure 4.2's attack path model. For example,
 A Bayesian approaches
 B Markov Modeling
 C Discrete Event System Specification (DEVS)

2 How are CSCs used in M&S for cyber defense? (Figure 4.8)

3 How do architectural constructs, subjects of M&S, form alternatives for strategic cyber decision making (Figure 4.8)?

4 Why is the McCumber model a better choice for developing cyber security scenarios (e.g. compared to Bell-LaPadula, Biba, Clark-Wilson, etc.)?

5 Who is the primary target customer for the MACE Taxonomy?

6 What are the key differences between Threat Models and Attack Scenarios?

7 Why is it important to differentiate between Cyber Effects and Military Activities in the MACE Taxonomy?
 A Are cyber effects always related to CIA?

8 How can the MACE Taxonomy be used in the standard threat modeling approaches (e.g. DREAD, STRIDE, etc.)?

5

Cyber Standards for Modeling and Simulation

The emerging cyber threat presents defense with military command and control issues of a type, scale, and scope not seen before (Stella Croom-Johnson 2016). Traditional processes have been stretched beyond their intended limits by the need to take into account not only the new factors and novel methods of attack introduced by the cyber threat but also the uncertainty regarding the efficiency of defensive countermeasures. In addition, on the INFOSEC "Hard Problems List" (Cyber Security and Information Assurance Interagency Working Group [CSIA IWG] 2006), under the heading "Information Provenance," identifies assuring the quality of shared data by tracking its evolution, as one of the most fundamental problems in information security (Dandurand and Serrano 2013).

While cyber standards are developing to facilitate common terminology and the efficiency of work, they are used for an array of applications, often disjoint; some examples are shown in Table 5.1.

Table 5.1 provides a sampling of cyber standards currently available, mostly for information assurance use. Traditional Modeling and Simulation (M&S) tools, however, were developed to address the questions surrounding conventional warfare, but not those surrounding the confidentiality, integrity, and availability of essential mission networks, or their respective impact on a scenario. In cyber space, some of these questions can be addressed by the use of models. These range from those that categorize cyber incidents using cyber information exchange standards (such as TAXII and CybOX) to tools that provide a structured expression of threat and attack (such as STIX), and those that provide a visual analysis, and subsequent Situational Awareness (SA) of candidate threat scenarios.

Models of this type are capable of translating diverse and constantly changing information into actionable knowledge, giving cyber defenders flexibility in their available responses, helping to understand cyber observables and incidents, and giving managers outside the cyber domain an improved awareness of how a given situation might develop. This area is developing very quickly, and new tools have often been developed to address a specific need, but without interoperability with

An Introduction to Cyber Modeling and Simulation, First Edition. Jerry M. Couretas.

Table 5.1 Example cyber standards.

Standard	Use	Author
Cyber Range Interoperability Standard (CRIS)	Connect cyber models on logical ranges for training exercises	Test Resource Management Center (TRMC)
Common Research Into Threats (CRITS)	Ability to communicate and share threats between organizations, government, and the public, opens up a more collaborative effort toward intelligence-based active threat defense (MITRE 2014).	MITRE
Common Vulnerability Specification System (CVSS)	Open framework for communicating the characteristics and impacts of IT vulnerabilities. Its quantitative model ensures repeatable accurate measurement while enabling users to see the underlying vulnerability characteristics that were used to generate the scores (NIST).	NIST
OCTAVE	The OCTAVE method is an approach used to assess an organization's information security needs. OCTAVE Allegro is the most recently developed and actively supported method. This method is based on two older versions called OCTAVE Original and OCTAVE-S (Carnegie Mellon University Software Engineering Institute [CMU SEI]).	CMU SEI
STIX/TAXII	Categorize cyber incidents using cyber information exchange standards	MITRE

other such tools being a primary consideration. There is a very real need for standardization not only of the structures and formats used by these tools, but also for a common language across all areas to reduce misunderstandings and to facilitate the speedy processing and dissemination of information.

There are a number of areas where M&S tools can contribute to the cyber defense effort and has considered some of the potential benefits that could be derived from the application of standards. Building on this and other work, this section will identify some of the tools and standards currently in use in these areas, highlighting the benefits that could be derived from the consistent application of standards, potentially including the introduction of a common language.

5.1 Cyber Modeling and Simulation Standards Background

The emerging cyber threat presents defense with military command and control issues of a type, scale, and scope not seen before. Traditional processes have been stretched beyond their intended limits by the need to take into account new factors, and novel methods, of attack introduced by the cyber threat. Traditional M&S tools were developed to address questions surrounding conventional warfare. However, they were not designed to address the questions surrounding the confidentiality, integrity, and availability of essential cyber components, which are needed to support missions at the network and mission layers, nor do they address the question of how to represent the impact on a conventional training scenario of the loss of one or more of the supporting elements. The first line of cyber defense is provided by monitoring the cyber events and observables triggered by potential threats at the different layers and ensuring that robust security configurations, practices, and components provide the optimum balance between protection and usability. However, it may be assumed that at some point attack vectors will succeed in penetrating the operational network. This means that military personnel in non-cyber roles need to be trained in how to recognize early indicators of potential cyberattacks and to understand the appropriate responses in such an eventuality.

For a holistic cyber approach to be truly effective, a systems approach is needed that embraces network defense, physical security, intelligence gathering, cyber response, operational training, and mission rehearsal. The main challenge in developing this approach is how to best integrate cyber standards with existing simulation standards to create a seamless representation of the impact of a cyberattack in non-cyber mission rehearsal and training exercises.

5.2 An Introduction to Cyber Standards for Modeling and Simulation

The breadth of the cyber domain makes a full survey of cyber tools and research an evolving pursuit; independent of M&S. MITRE, for example, has done extensive work on cyber description, providing tools for both cyber specialists and more traditional operators who rely on cyber systems. In addition, M&S-based approaches, including Tolk's hierarchy (Tolk and Muguira 2003), provide a level of abstraction that helps capture the sometimes opaque cyber system relations.

5.2.1 MITRE's (MITRE) Cyber Threat Information Standards

CybOX™, STIX™, and TAXII™ were developed by the MITRE Corporation as part of an initiative by the US Department of Homeland Security (DHS) Office of Cybersecurity and Communications, National Cybersecurity and Communications Integration Center (NCCIC). The aim was to automate and structure operational cybersecurity information-sharing techniques across the globe, but even from the brief summary of these standards given below, synergies with the SISO objectives and standards are clearly apparent. These standards have transitioned to OASIS (OASIS) and are now open standards (Table 5.2).

Table 5.2 Cyber description tools (MITRE).

Title	Description
CybOX™ (Cyber Observable eXpression) (MITRE)	The MITRE website summarizes CybOX™ as "a standardized language for encoding and communicating high-fidelity information about cyber observables." It offers a common structure at the enterprise level that can be used to represent dynamic events and static attributes in the network of interest, together with the associated corrective actions taken.
STIX™ (Structured Threat Information eXpression) (MITRE)	The STIX™ framework uses an XML schema to express cyber threat information with a view to enabling the sharing of that information and generating a cyber threat analysis language. It tries to build up the language by using referential relations between tables and nodes, with the goal of creating a standardized way of representing the cyber threat.
TAXII™ (Trusted Automated eXchange of Indicator Information) (MITRE)	TAXII™ is a standardized way of defining a set of services and message exchanges for exchanging cyber threat information. It uses XML and is service-oriented with four options (Inbox, Poll, Collection Management, and Discovery) and three sharing models (Hub and Spoke, Source/Subscriber, and Peer to Peer).

CybOX™, TAXII™, and many other information assurance tools fall into the "cyber for cyber" category, and are primarily tools which allow cyber professionals to communicate with each other. STIX™ is also a "cyber for cyber" tool: part of its functionality is as a common language for information sharing, but it has additional relevance to this section as it also provides a model for simulations to represent the different types of attacks.

These standards are focused on enabling information sharing between cyber defense tools and leveraging them for simulation requires a differentiation to be made between the two different communities:

1) Cyber for Cyber (C4C)
2) Cyber for Others (C4O)

The differences between the two types of simulation could be summarized as simulations for C4C personnel being task oriented and covering the tools, techniques, and procedures used in cyber defense. Those for C4O personnel are impact oriented and facilitate consideration of the measures needed to minimize the effects of a cyberattack on a mission. In this context, C4C tools address capability training and C4O tools address awareness.

Figure 5.1 goes some way to illustrating this difference, while highlighting the overlap between items of interest to non-cyber military personnel (C4O) and those of interest to cyber operators (C4C) within the military. However,

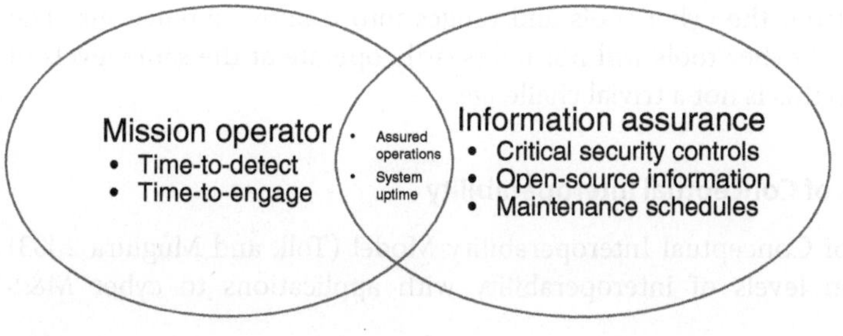

Figure 5.1 Military operator and cyber IA overlaps.

even though M&S systems and operator training systems are composed of the same underlying technologies, the policy and training overlap between them is limited.

It is only recently that attempts have been made to integrate the cyber element with the mission rehearsal and training simulations for C4O personnel. Due to the way the term "cyber" has been overloaded, it is currently a challenge to differentiate between the training for the respective personnel shown in Figure 5.1. For example, CyberCiege (Thompson and Irvine 2011), a well-known tool within the cyber community, is more likely to fit into the Cyber Information Assurance sphere in Figure 5.1, even though it is supplied by the US Naval Postgraduate School which would at first sight make it seem a fit for Military Operators. In addition, the Cyber Operational Architecture Training System (COATS) (Wells and Bryan 2015; Morse et al. 2014a, b) described below, although currently aspirational and described in terms of distributed simulation (i.e. information technology) terms, is intended as more of a training tool.

5.2.2 Cyber Operational Architecture Training System

The US COATS (Wells and Bryan 2015) project examined how a general-purpose cyber effects data model might allow the outputs of a cyber range to be injected into a simulation similar to those used in mission rehearsal and training exercises for non-cyber military personnel. This used the IEEE standard 1730-2010™ (IEEE recommended Practice for Distributed Simulation Engineering and Execution Process [DSEEP]) (IEEE Std 1730-2010) to support the integration of simulations for the two groups and to support the creation of linkages between them. A natural next stage is the integration of multiple cyber ranges and tools with multiple simulations (Damodaran and Couretas 2015), currently called a logical range. Implicit in this activity is the need for standardized formats, semantics, and architectures to enable interoperability between cyber ranges and tools, as well as between cyber ranges and tools and simulation tools. An additional issue is the need to reconcile the issues involved with porting data from the cyber tools and ranges into training simulations. The simulations and cyber tools will not necessarily operate at the same levels of abstraction, so this is not a trivial challenge.

5.2.3 Levels of Conceptual Interoperability

Tolk's levels of Conceptual Interoperability Model (Tolk and Muguira 2003) outlines seven levels of interoperability with applications to cyber M&S (Figure 5.2).

Tolk's levels span a range that goes from stand-alone tools, with no interoperability, to tools with conceptual interoperability where all elements of a model are designed with interoperability being a key requirement taken into account from their design stage.

From the Tolk model, it can be inferred that cyber tools used by C4C personnel, especially ranges, have traditionally been designed at Level 0 (emulators, no interoperability) to ensure maximum security both from incoming cyberattacks and accidental (outgoing) data leakage. As cyber ranges mostly use emulators, they tend to be confined to using real (or virtualized) Computer Information Systems (CIS) components or protocols, and have relatively few constructive components. This is because constructive simulations are software and may be adapted easily to implement any interface for interoperability: cyber range components, as given above, are mostly CIS components (live or

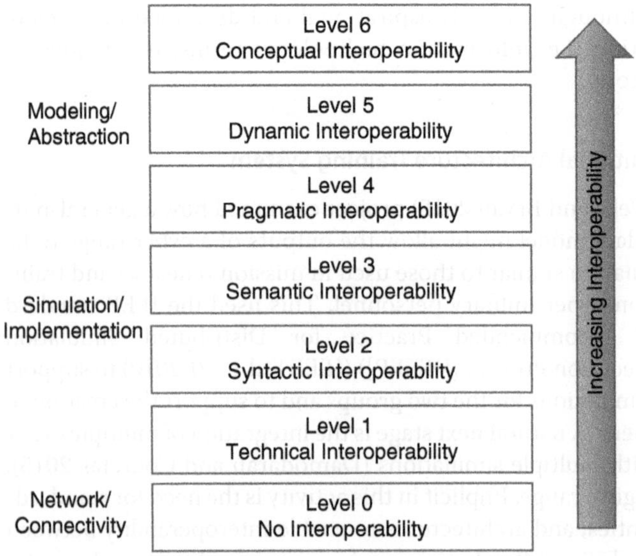

Figure 5.2 Levels of Conceptual Interoperability Model.

virtual), and it is more difficult to modify them to provide functional control interfaces. On the other hand, many C4C tools include scenarios requiring CIS components rather than software implementations.

While this is now starting to change, the standards for cyber tools are predominantly technical standards focused at Levels 0–3, whereas SISO standards are more focused at Levels 4 and 5. Simulations used by C4O personnel in training and mission rehearsal exercises for non-cyber military personnel often include constructive elements (e.g. Computer Generated Forces [CGFs]) making it easier for them to reach these higher levels of interoperability.

At this stage it should be highlighted that an emerging finding from the COATS work was the importance of ensuring a common understanding of the vocabulary used. When working across boundaries between federated training simulations and tools in the cyber domain, each community has its own vocabulary and the meaning of a given word can differ substantially according to context. Many words are common between the cyber and simulation communities, but this can give rise to confusion when working with members of both communities. For instance, a representative from the simulation world (C4O) might draw their interpretation from the mission perspective and understand "synthetic environment" to mean terrain or CGFs. A representative from the cyber world (C4C) is more likely to draw their interpretation from a technical perspective and, in the same conversation, understand "synthetic environment" to mean a cyber range. It is important for all users to have the same understanding of what is meant in the given context and these two perspectives will need to be reconciled and aligned before work can start to create a common vocabulary.

5.3 Standards Overview – Cyber vs. Simulation

This section will take a brief look at the standards used in both the cyber domain and the simulation domain. It will show how standards are a key element in each domain, and how they are aimed at different levels of Tolk's Conceptual Interoperability model.

From even a brief comparison it becomes apparent that most standards used in the cyber community operate at Levels 0–3 of the Tolk model. These are focused on ensuring the confidentiality, integrity, and availability of data and assume a relatively low level of trust between organizations and nations. SISO standards typically address the physical layer, and operate at Levels 4 and 5 of the Tolk model. They are aimed at sharing data between federations of simulations.

Cyber-specific training is currently intended to assure that operators can recover from a cyberattack. The models used are often attack models, with the systems holding no representation of the associated defense models. As a consequence, although cyber ranges use metrics to measure defensive capabilities, these only reflect how successfully a network is defended. They do not reflect whether the relevant countermeasures were successful in preventing attackers from attaining their objectives. Part of this will depend on whether the affected non-cyber personnel were aware of the appropriate and proportionate reactions – and behaved accordingly.

5.3.1 Simulation Interoperability Standards Organization (SISO) Standards

This section is not a comprehensive review of SISO standards but gives a brief overview of those deemed to be of most interest in the cyber context.

5.3.1.1 C2SIM: Command and Control Systems – Simulation Systems Interoperation

The C2SIM Product Development Group (PDG) and Product Support Group (PSG) are an evolution of SISO groups that developed the Standard for Military Scenario Definition Language SISO-STD-007-2008 (MSDL) (SISO-STD-007-2008) and SISO-STD-011-2014 Standard for Coalition Battle Management Language (C-BML) Phase 1 (SISO-STD-011-2014). MSDL is a standardized XML-based language that enables the sharing of scenario data between synthetic environments and the C4I elements of a simulation. C-BML is an XML-based language to express commanders' intent, and is structured in such a way that it can send commands and receive reports across a combination of command and control (C2) systems, live virtual and constructive (LVC) M&S systems, and autonomous systems. It is primarily focused on simulations running at Levels 4 and5 of the Tolk model.

The C2 element of C2SIM gives it the potential to support the representation of a cyberattack. Not only could it highlight when a degradation of communication might be an indicator of possible cyberattack but could also facilitate the representation of the impact of packets failing to arrive, or packet interception with dissemination of false information and spurious commands arising from their subsequent onward transmission. Current standards are challenged to make specific provision for such a representation of the impact of a cyberattack. The implementation of this is likely to be a complex task, probably needing cyber components to be mapped to missions, and will need to be addressed from the dual perspective of both of mission SA, and of Course of Action analysis.

5.3.1.2 DSEEP: IEEE Standard 1730-210™ (IEEE Recommended Practice for Distributed Simulation Engineering and Execution Process) (IEEE Std 1730-2010)

The DSEEP process was developed and is maintained by SISO. It defines a seven-step process that can be deconstructed into component tasks and activities that set out best practice for the design, development, integration, and testing of simulation environments. The normal diagrammatic representation shows seven sequential steps running from the definition of objectives through to the final stage of after-action review and analysis of results. In practice, the development sequence follows an iterative spiral model rather than a waterfall model, with any given stage in the process having the potential to generate a need to revisit and adjust the outputs of earlier stages.

In traditional exercises, the level of abstraction at which many simulations operate means that the impact (rather than the reality) of a cyberattack needs to be represented. These impacts, such as loss of power, are often no different to the effects experienced from a traditional, kinetic attack. Therefore, many of these effects can already be represented in simulations and the use of DSEEP would help to integrate a representation of cyber into the more traditional scenarios.

A number of overlays exist to tailor DSEEP for specific circumstances, but the process currently makes no specific provision for cyber. The DSEEP documentation sets out a detailed product flow for each of the seven steps, breaking them down into component activities. Subsequent sections provide more detail about each activity, suggesting inputs, recommended tasks, and outcomes. Existing overlays outline where their activities are identical with the baseline DSEEP and offer guidance about how to manage differences between the overlay and the generic DSEEP activity descriptions. An overlay for cyber that did this would facilitate the integration of the representation of a cyberattack into mission rehearsal and training exercises for non-cyber military personnel.

5.3.1.3 DIS: IEEE Standard 1278™ Series, "IEEE Standard for Distributed Interactive Simulation" (DIS) (IEEE Std 1278 Series)

The NATO Allied Modeling & Simulation Standards Profile (AMSP) (NATO 2015) states that "DIS is a protocol for linking simulations of various types at multiple locations to create realistic, complex, virtual worlds for the simulation of highly interactive activities." An important attribute of this standard is that it facilitates interoperability between systems designed to achieve different objectives, with structures, format, and language suited to their own objectives. Exercises using DIS are intended to support a mixture of virtual entities with computer-controlled behavior (CGFs), virtual entities with live operators (human-in-the-loop simulators), live entities (operational platforms and test and evaluation systems), and constructive entities (war games and other automated simulations).

Federations using DIS are relatively simple to establish, but a number of factors make it unlikely to be the best base architecture for the large mission rehearsal and training simulations used in training C4O personnel. However, many legacy systems, and some systems from other nations – including many cyber tools and ranges – have been designed to operate using DIS.

This should not be seen as an insuperable barrier to their inclusion in the larger exercises, as a suitable gateway can be used to connect them with one or more cyber effects models similar to those suggested by the COATS project.

5.3.1.4 HLA-E: IEEE Std 1516™, High-Level Architecture for M&S (HLA) (IEEE Std 1516)

The AMSP (NATO 2015) states that HLA-E "was developed to provide a common technical architecture that facilitates the reuse and interoperation of simulation systems and assets. It provides a general framework within which developers can structure and describe their simulation systems and/or assets and interoperate with other simulation systems and assets." Each federation agreement uses a Federation Object Model (FOM) to specify the information to be exchanged by federates at run time. This defines the couplings that will take place between federates allowing participants to know what data they can expect to receive, and the format of the data. For a federation to include a specific representation of cyber, a FOM would need to include descriptions of the disruption a cyberattack would cause to the interactions within the simulation.

One well-known example of a FOM is the NATO NETN (NATO Education and Training Network) FOM. This takes a modular approach to defining the interactions, one module dealing with those between the C2 and the simulation. This could potentially be extended to facilitate a representation of the impact of a cyberattack in C4O simulations. At this stage it is not possible to ascertain whether there is a need for a separate cyber-specific FOM.

5.3.2 Cyber Standards

In contrast to the SISO standards, cyber standards (Levels 1–3 of the Tolk model) are aimed at ensuring the secure transmission of data packets between network nodes, and the protection of a network from unauthorized activity, including the introduction of threat vectors. Figure 5.3 illustrates how different controls and countermeasures are appropriate to the different points along the attack path, and – should these not be addressed – the potential consequences, with the effects of those consequences.

Although Figure 5.3's 'Bow-Tie' is a cyber construct, it has a wider application within the context of this section, as it shows how differing but equally valid interpretations are possible at both the cyber and simulation levels. A seamless progression in the representation of cyber as we move through the levels of the Tolk model may well be possible, provided there is a clear understanding of the context and objectives of what needs to be represented for events at each level. At the cyber level, the event represents an attack entering

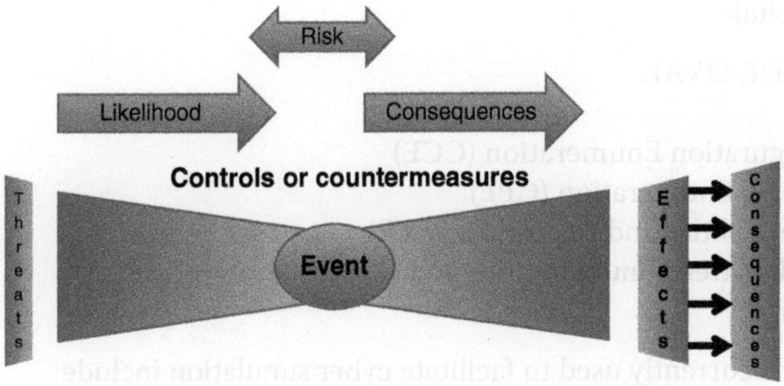

Figure 5.3 Cyber "Bow-Tie" – Prevention, Attack, and Remediation

a network, with the effects and consequences being those manifested at the network level. In the context of simulations used for mission rehearsal and training, the event would be the compromise of a network (irrespective of the cause) with the effects and consequences highlighted being those on the mission rather than on the network. This would make the Bow-Tie a powerful enabler for a dialogue about cyber tools and their possible contribution to situational awareness.

In addition to leveraging threat information standards, the cyber domain uses them to help manage both pre-event sanitization and post-event remediation. For example, each of the SANS 20 Controls is designed to provide automatic, machine programmable approaches to respond to cyber threats. Their stated goal is to "protect critical assets, infrastructure, and information by strengthening your organization's defensive posture through continuous, automated protection and monitoring of your sensitive information technology infrastructure to reduce compromises, minimize the need for recovery efforts, and lower associated costs." These are paralleled in Australia by the Australian Signal Directorate's "Top 4" preventive techniques, said to prevent 85% of attacks. In addition, the Security Content Automation Protocol (SCAP) includes specifications for organizing and expressing security-related information in standardized ways, as well as related reference data such as unique identifiers for vulnerabilities:

- Languages:
  - Asset Reporting Format (ARF)
  - Open Checklist Interactive Language (OCIL)
  - Open Checklist Reporting language (OCRL)

- Measures:
  - Common Configuration Scoring System (CCSS)
  - Common Misuse Scoring System (CMSS)

SCAP affiliations include:

- Languages: XSCCDF, OVAL
- Enumerations:
  - Common Configuration Enumeration (CCE)
  - Common Platform Enumeration (CPE)
  - Common Vulnerabilities and Exposures (CVE)
  - Common Attack Pattern Enumerations and Classifications (CAPEC)
- Measures: (CVSS)

In addition, standards currently used to facilitate cyber simulation include:

- RFC 5070
- STIX™
- 8500.01
- NIST SCAP
- NATO NISP
- STANAG 4631, 5067
- IO PDU from the IEEE Std 1278™ series, "IEEE Standard for Distributed Interactive Simulation" (DIS)

The heavy reliance on standards of cyber ranges makes them a natural fit for cyber modeling and simulation. Also, while Logical Ranges (Damodaran and Couretas 2015) (multiple federated ranges) show promise for Levels 2 through 6 of Tolk's model, the continued development of formal standards will help the training community to adopt this concept in a coherent and coordinated way.

Within the cyber community, standards that result in ranges and other cyber tools sharing data are emerging, but are not yet in general use. The NATO Allied Modeling & Simulation Standards Profile (AMSP) makes no reference to the network level standards most commonly used by cyber tools. The OASIS standards (OASIS) show that the cyber community is starting to consider interoperability issues, although this is still in the early stages. However, these standards appear to have evolved without the objective of integration with SISO standards and make no reference to them. In turn, the SISO standards currently have no specific provision for the inclusion of a cyber element in a simulation.

5.4 Conclusions

In order to achieve true all-round cyber SA, the information from cyber tools and ranges needs to be exploitable by other cyber tools and ranges, as well as by the simulations used in mission rehearsal and training exercises for C4O (non-cyber specialists). The previous sections have shown how, although tools in each category might comply with the standards used with in its own area, it is unusual for their design to include provision for interoperability outside their own domain.

Benefits that could be derived from the consistent introduction (and use) of standards to achieve interoperability between the two sides of the equation include:

- Improved general cyber awareness across cyber and non-cyber military personnel alike.
- Improved awareness among non-cyber military personnel of how to recognize a potential cyberattack, and of how to respond to such an event appropriately and proportionately.
- Improved awareness of the potential impact of a cyberattack on an operational mission – for cyber and non-cyber personnel alike.
- An improved ability to conduct "What If" and Course of Action analyses on scenarios that include a cyber element.
- Improved awareness among cyber personnel of the potential for their activities to enhance non-cyber mission rehearsal and training exercises.

The interoperability question is not confined to technology tools, but extends to the vocabulary used by C4O and C4C personnel. Different interpretation of words that are at first sight common between the communities can have an unexpected on situational awareness. A common, standardized language would help to reduce misunderstandings, and lead to an improved common understanding and situational awareness. Enhanced mutual awareness of these respective standards increases the likelihood that tools in both areas would be designed and used bearing in mind interoperability with the others.

5.5 Questions

1 What advantages will standards for cyber M&S provide?

2 Why is the C4C/C4O differentiation important for cyber M&S?

3 What cyber information types is Tolk's Hierarchy best suited for?

4 How do attack models translate between the training and testing domains?

6

Cyber Course of Action (COA) Strategies

This chapter examines how decision-making in cyber defense may benefit from Modeling and Simulation (M&S). The cyber domain presents scale and scope issues that require decision aids to meet the accuracy and timeliness demands for securing the network. The use of "models," for cyber decision support spans from longer-term decision support, in categorizing projected network events, to real-time visualization of developing threats, and using these models to analyze attack graphs and projected second- and third-order effects for each COA candidate.

Developing COAs to respond to cyberattacks is especially challenging with the rise of threat capability, and the number of nefarious actors (Mandiant 2014). Cyber actors have the ability to coordinate (e.g. via botnets [Kotenko 2005]) and scale an attack at time constants potentially much faster than standard human cognition. M&S, in Decision Support Systems (DSS), can enhance situational awareness (SA) through training. The knowledge imparted by M&S, used in the design and development of DSS, trades directly against the technical advantages and experience of a cyber attacker. Understanding how a DSS' COA effectiveness will be measured is therefore key in DSS design.[1]

6.1 Cyber Course of Action (COA) Background

6.1.1 Effects-Based Cyber-COA Optimization Technology and Experiments (EBCOTE) Project

In 2004, DARPA developed a cyber test bed for real-time evaluation of COA impact, evaluating performance and effectiveness, for time-critical targeting systems (Defense Advanced Research Projects Agency [DARPA] 2004). A

1 While a number of cyber decision support systems are currently advertised (e.g. Cytegic (http://cytegic.com/cdss/), Panoptesec (http://www.panoptesec.eu/), …,) the focus here is on the use of M&S to produce cyber defense DSSs.

An Introduction to Cyber Modeling and Simulation, First Edition. Jerry M. Couretas.

fundamental challenge for modern Battle Management/Command, Control, and Communications (BMC3) systems is to withstand attacks against their constituent computer and communication subsystems. However, measures to safeguard or respond to a cyberattack against a BMC3 system invariably disrupt the processing flow within that system. Thus, disruptions may ultimately affect mission effectiveness, and a prudent strategy is to predict those impacts before committing to a specific response or safeguard.

EBCOTE studied the problem of quality of service (QoS) assurance in BMC3 systems in the context of a Time Critical Targeting (TCT) scenario, focusing on the mission as a workflow, and determining mission effectiveness based on how the degradation of the underlying IT system affected the mission.

As shown in Figure 6.1, the three research phases of EBCOTE included both offline and on-line evaluation of mission impact due to BMC3 workflow disruption, including on-line generation/optimization of cyber COAs.

Figure 6.1's EBCOTE, an early mission modeling success, was expanded on with the broader mission evaluation capability of the "Analyzing Mission Impacts of Cyber Actions" (AMICA) prototype (Noel et al. 2015), more recently promoted by MITRE.

6.1.2 Crown Jewels Analysis

In addition to AMICA, MITRE developed "Crown Jewels Analysis" (CJA) (MITRE), a process for identifying the cyber assets most critical to the accomplishment of an organization's mission. CJA is also an informal name for

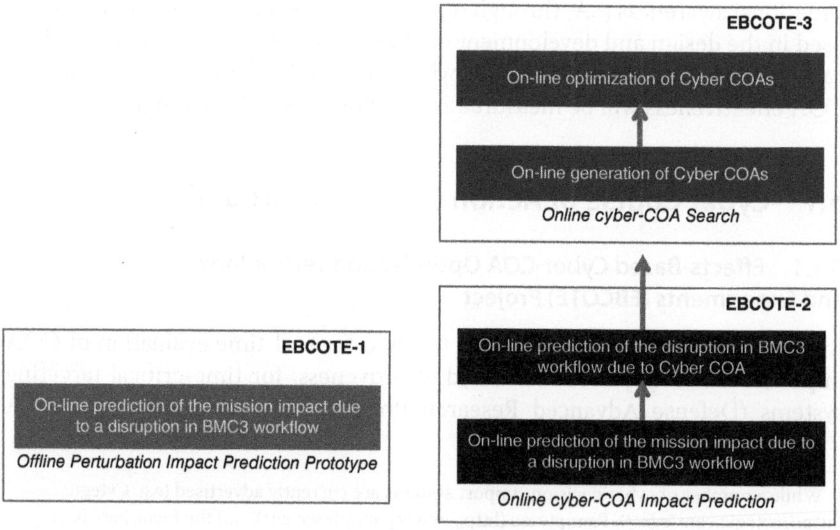

Figure 6.1 Three research phases in the evolution of the EBCOTE system.

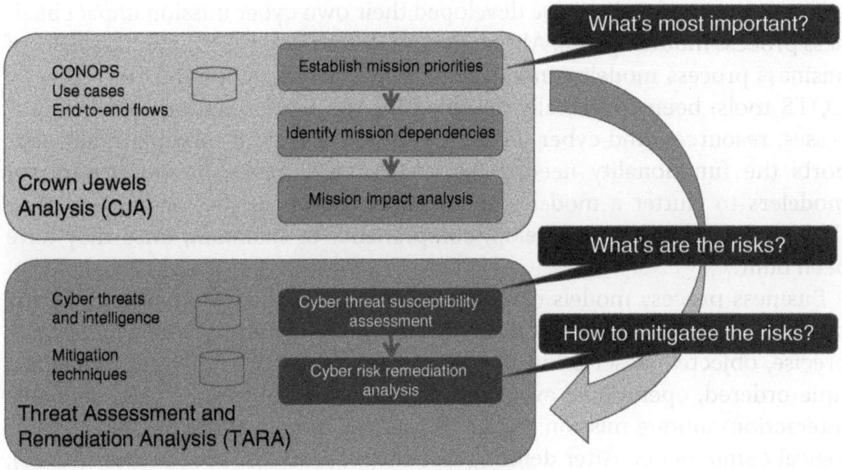

Figure 6.2 The Mission Assurance Engineering (MAE) Process.

Mission-Based Critical Information Technology (IT) Asset Identification. It is a subset of broader analyses that identify all types of mission-critical assets (Figure 6.2).

Mission Assurance Engineering (MAE) offers a common, repeatable, risk management process that is part of building secure and resilient systems. The underlying premise for performing a CJA, as part of the MAE, is that protection strategies focused entirely on "keeping the adversary out" are challenged by advanced cyber attackers; requiring defenders to maintain vigilance through processes like MAE, informed by a periodic CJA. Because it is difficult and costly to design every component of a system to be hardened against all conceivable attacks, a CJA helps identify the most important cyber assets to an organization's mission – providing a baseline for systems engineers, designers, and operators to focus on, to ensure that these critical components are secure.

6.1.3 Cyber Mission Impact Assessment (CMIA) Tool

In addition to CJA, the Cyber Mission Impact Assessment (CMIA) is one approach for performing a cyber mission risk assessment (Musman et al. 2013). From a systems engineering perspective, CMIA makes it possible to perform system assessments by simulating the application of potential security and resilience methods to a system within the mission context. Since effective resilience methods either prevent or mitigate the impacts of cyber incidents, when combined with the probability that bad events will occur, the impacts computed by CMIA address the "amount of loss" part of the risk equation. The CMIA tool extensions include combining it with a topological attack model to support mission assurance assessments and return-on-investment calculations.

The creators of CMIA have developed their own cyber mission impact business process modeling tool. Although it implements only a functional subset of business process modeling notation (BPMN), it has, unlike the more generic COTS tools, been specifically designed for the representation of cyber processes, resources, and cyber incident effects. As such, it more naturally supports the functionality needed for CMIAs and makes it unnecessary for modelers to clutter a model with extraneous content that ends up making those models harder to develop, comprehend, or maintain, once they have been built.

Business process models can be used to represent mission systems in the context of their execution of mission threads. A mission thread represents a precise, objective description of a task. In other words, a mission thread is a time-ordered, operational event description that captures discrete, definable interactions among mission resources, such as human operators and technological components. After defining testable measures of effectiveness (MOE), measures of performance (MOP), and key performance parameters (KPP) for the modeled mission, the process model captures how the performance of mission activities contributes to achieving them (Figure 6.3).

Since a process model captures activity, control, and information flows, it is possible to evaluate alternate variants of resource assignments, information, and control flows in order to assess potential architecture improvements. For example, flexibly increasing nodal capacity, by leveraging the cloud (Hariri et al. 2003), might improve resiliency for defensive scenarios. Another example might be to develop an MOP for trained operators increasing consistency in security practices, ensuring good housekeeping keeps the local network in good working order.

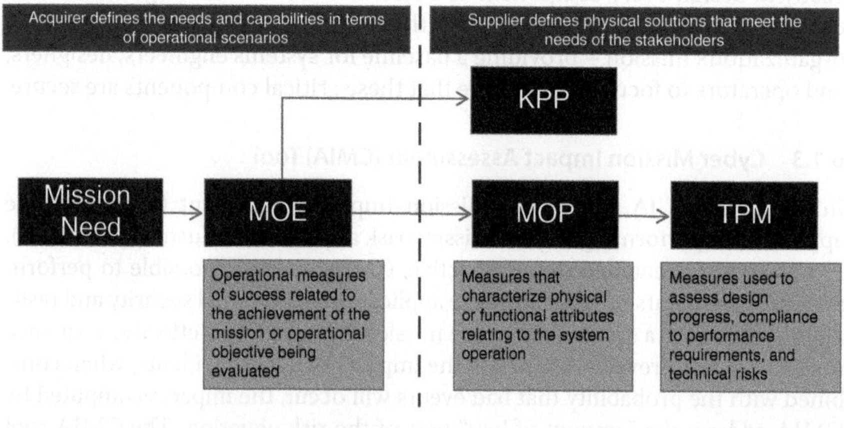

Figure 6.3 Mission needs, MOPs, MOEs, and KPPs.

6.1.4 Analyzing Mission Impacts of Cyber Actions

Analyzing Mission Impacts of Cyber Actions (AMICA) provides an approach for understanding the broader mission impacts of cyberattacks (MITRE 2015; Noel et al. 2015). AMICA combines process modeling, discrete-event simulation, graph-based dependency modeling, and dynamic visualizations. This is a novel convergence of process modeling/simulation and automated attack graph generation.

6.1.4.1 AMICA and Process Modeling

As shown in Figure 6.4, AMICA, similar to EBCOTE, models a mission and includes the respective mission entity IT dependencies via a multilayered architecture. Network mapping tools (e.g. NMAP, Nessus, etc.) are used in the real world to validate the dependencies between the respective mission nodes and supporting IT infrastructure.

AMICA's ability to evaluate COA possibilities is similar to the EBCOTE work in evaluating mission impacts due to supporting IT anomalies.

6.1.4.2 AMICA and Attack Graphs

Attack graphs focus purely on the supporting IT layer's nodes, as described in Figure 6.4. One of the goals of using attack graph analysis is to find the "reachability," or potentially vulnerable connections, in the supporting IT system. A formal attack graph is defined in Equation (6.1) (Wang et al. 2006; Jajodia et al. 2015).

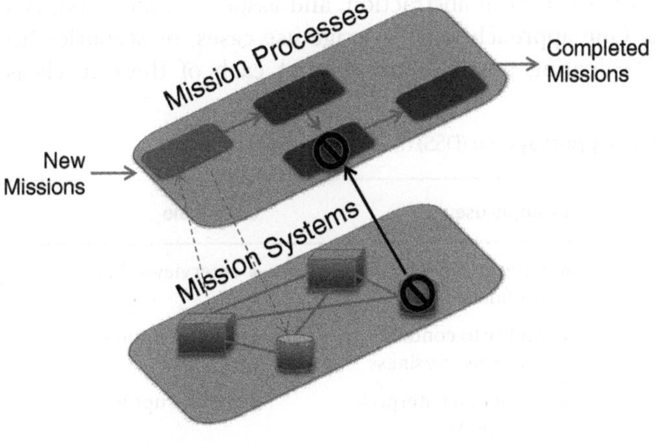

Figure 6.4 AMICA – Information Technology (IT) to mission simulator.

As shown in Figure 6.4, mission processes are dependent on underlying mission systems, often in the form of a computer system that is vulnerable to cyber attack. AMICA therefore provides the mission impact, often in the form of availability, due to an underlying system's cyber compromise.

6.2 Cyber Defense Measurables – Decision Support System (DSS) Evaluation Criteria

While Cyber "Observe Orient Decide Act" (OODA) loops (Gagnon et al. 2010) have been explored in previous research, identifying the observables remains a challenge in DSS development. For example, the cyber defender, for whom the DSS is designed to serve, is tasked with securing the network over a range of performance metrics, for a given attack scenario. If the threat is unknown, or in a large system, and the impact of a certain COA is uncertain, M&S contributes by looking at the impact of an attack and determining mitigation strategies from there. As shown in Table 6.1, COAs can leverage Table 6.1's metrics in determining how to evaluate models at different abstraction levels.

Policy Effectiveness, somewhat novel to a technical audience, is an approach that has worked for industrial, automobile, and aircraft safety (Economist). In addition, Table 6.1 shows that there are a variety of approaches for evaluating a given system, based on the level of decision making the DSS is designed to provide. DSS COAs have the somewhat conflicting requirements of being fast, accurate, and current. Meeting these objectives implies focusing respective DSSs on the right abstraction level, measured in the right way (Table 6.1).

Determining the correct level of abstraction, and associated metric(s), is a recurring challenge. One approach is to leverage use cases, or scenarios, to evaluate DSS use. Evaluating possible scenarios at each of these levels is

Table 6.1 Cyber Decision Support System (DSS) metrics and example use.

Metric	Example use	Collectible
Measure of Policy Effectiveness (MOPE)	• Enterprise perception/ confidence	• Interviews (human)
Measure of Enterprise Effectiveness (MOEE)	• Ability to conduct enterprise business	• System uptime
Measure of System Effectiveness (MOSE)	• Impact on enterprise operations	• System uptime
Measure of Performance (MOP)	• Response time • System availability	• Time to detect/respond

amenable to "Data Farming," or design of experiments (DOE), for rapid prototyping for different possible scenarios.

As shown in Table 6.1, system metrics provide the DSS developer an opportunity to specify, or evaluate, system performance, in terms of measurables, during the project's design phase. Other considerations during the DSS' design include the cognition enhancements that the system will provide. For example, cyber SA is commonly denominated in "network events," activity monitored by Security Information and Event Management (SIEM) tools, thereby distilled to a current system state representation.

6.2.1 Visual Analytics

In modern defense and security operations, analysts are faced with ever-growing data sets, widely scoped, that cause significant information overload problems and prevent good situation awareness. Visual Analytics (VA) is a way of handling and making sense of massive data sets by using interactive visualization technologies and human cognitive abilities. Defense R&D Canada conducted a review (Lavigne and Gouin 2014) of the applicability of VA to support military and security operations.

6.2.1.1 Network Data – Basic Units of Decision Support Systems (DSSs)

Network events, data centric, the common currency of monitoring and control, are the basis for SIEM tools. While SIEM can be comprehensive, by including insider threat integration (Callahan 2013), the scale of current IT systems (i.e. the number and variety of events) makes event monitoring for administrators a "Big Data" issue (Grimaila et al. 2012).

6.2.1.2 Big Data for Cyber DSS Development and Evaluation

While SIEM tools naturally use large data sets collected on a system of interest, the broader concept of quality stems from focused application of data resources to manufacturing. For example, the basics of data evaluation, popularized by quality control (Deming 1967; Deming 2010), provides a basis for the collection and manipulation of evidence for cyber domain decision making[2]; as it does for other domains. The application of Deming's work, popularized by the success of the Japanese auto industry, led to the revolutionizing of both the auto industry and manufacturing in general.

While Deming's use of data for manufacturing quality management is well known, now, this was a very complicated subject only a few decades ago, when computers were first being introduced to front offices and the factory floor.

2 Additional contextual uses of data, leveraging high-frequency trading algorithms used in the stock market, for example (Lo and Hasanhodzic 2010), provide potential case studies for cyber DSS development with large data sets (e.g. log/flow data).

Similar to contemporary "cyber," getting a handle on "quality" was a seemingly qualitative and semi-subjective pursuit, requiring the development of policies and processes to solidify lessons learned.[3] This innovative use of computers, and data, led to unprecedented improvements across society.

6.2.1.3 COMPSTAT and People Data

In addition to manufacturing quality, early successes of "big data" include COMPSTAT (Henry 2002), a tool that law enforcement uses to evaluate urban zones for specific crime increases, with the idea of prescribing the right kinds of policing via data-based decision making. COMPSTAT's use of "hot spots" to direct emergency responders, led to geometric crime decreases in New York, Chicago, Los Angeles, and Washington DC immediately after adoption in the 1990s.

While cluster evaluation is common to data analysis, "hot spots" give an additional frequency, recency, and likelihood view of empirical data. In the case of cyber, this information is log data, describing legitimate and nefarious user interaction with the system of interest. COMPSTAT's lesson for cyber practitioners is the mixing of policy with technical collection, sometimes a challenge in the computer science-centric community of cyber practitioners.

Leveraging empirical data for DSS development complements cyber defense via ease of on-line data collection. In addition, due to the widely scoped nature of cyber threats, both open-source centers and information analysis centers are popular data sources. An example commercial cyber DSS that spans from data collection to intelligence reporting is provided in Figure 6.5.

Figure 6.5's DSS will require continual updating and leveraging of cyber counterintelligence capabilities (Duvenage and von Solms 2013) to stay current with the threat. Industry publications, examples of which are shown in

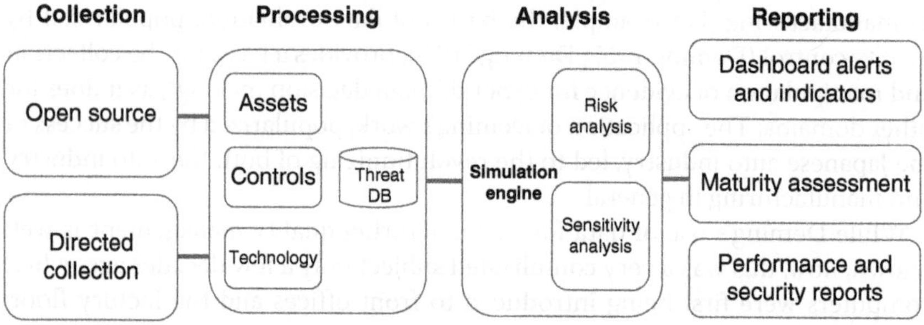

Figure 6.5 Cyber decision support system.

3 Quality vs. Security is one of the subjects addressed in Gollmann et al. (2006).

Section 12.1.3, provide valuable updates concerning the cyber threat, and how it may be changing. Understanding the data is a challenge – visualization may provide insight into relationships in threat data.

6.2.2 Managing Cyber Events

Figure 6.5's longer-term system evaluation is complemented by leveraging the "hot spot" multidimensional description of the event. This is a near real-time approach for enabling response teams to understand as much information as possible when responding to an event. In addition, these data sets can be used as an evaluation system for understanding progress of the remediation COA. Longer-term analysis will require more advanced data evaluation, such as data farming.

6.2.2.1 Data Farming for Cyber

"Data Farming," as used by the NATO Modeling and Simulation Group (NMSG)-124, is another option for the examination of multiple variants of a scenario within a very short time frame; along with semi-automatic analysis of extensive simulation experiments. Data Farming (Choo et al. 2008), one example for rapidly evaluating many possible alternatives, can be applied to:

- quantitative analysis of complex questions
- sensitivity studies
- stable system states and their transitions
- creating a "Big Picture" solution landscape
- enabling "what if" analyses
- gaining robust results

Valid models provide opportunities to anticipate the impacts of alternative COAs, comparing them before taking action – providing the "best" COA for a given scenario. The result may be a dynamic checklist showing different COAs and the expected gains (quality of mitigation) and losses (limitation of services).

In addition to using Data Farming to experiment with the possible COAs, the next natural question is how to understand the accuracy and bounds for a particular DSS. Given that a DSS is a software system, and we believe that indigenous system risk can be captured with PPPT (Chapter 10). Verification, Validation, and Accreditation (VV&A) is one approach for ensuring that a system meets its designed, and intended, use. Performing cyber DSS COA VV&A, while aspirational at this point due to the limited empirical data sets on both the threat and associated responses, is available via the recent General Methodology for Verification and Validation (GM-VV) (Roza et al. 2013) for estimating the confidence level in candidate COAs.

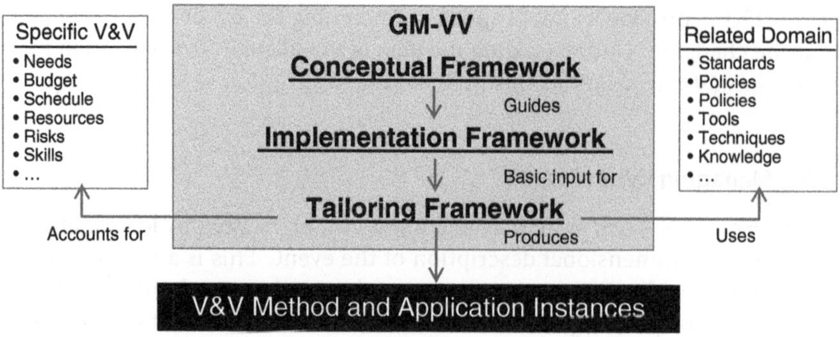

Figure 6.6 General Methodology for Verification and Validation (GM-VV) technical framework design and operational use concept.

6.2.3 DSS COA and VV&A

While COA simulations range from actual botnet evaluations (Kotenko 2005) to team-based training, it is a useful exercise to determine, at the beginning of a DSS design, both how we will be evaluating the DSS and the boundaries of its intended use. Formal processes (Roza et al. 2013) exist to ensure that a system meets its intended use (Figure 6.6).

Figure 6.6 provides a standard approach for implementing VV&A on a DSS and its respective COAs. More general, M&S-based approaches (Zeigler and Nutaro 2016) add flexibility that may be required in the cyber domain. Use of the GM-VV will likely provide leadership the confidence that a given cyber defense measure has been clearly thought through.

6.3 Cyber Situational Awareness (SA)

While long considered an important aspect of strategic and theatre planning, SA is the linchpin for both cyber planning and execution. As stated in Joint Doctrine (Joint Chiefs of Staff 2014), before military activities in the information environment can be accurately and effectively planned, the "state" of the environment must be understood (Robinson and Cybenko 2012). At its core, cyber SA requires understanding the environment in terms of how information, events, and actions will impact goals and objectives, both now and in the near future. Joint Information Operations (IO) doctrine defines the three layers of information as the physical, informational, and cognitive, as shown in Figure 6.7 (Joint Chiefs of Staff 2014).

The majority of cyber work is currently focused on the physical and informational aspects of the network to inform cyber SA. This includes leveraging

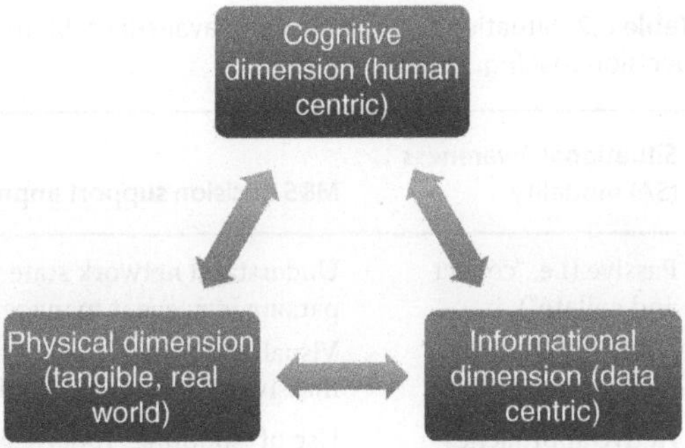

Figure 6.7 Three layers of Information Operations – physical, informational, and cognitive.

available network data for time saving, and high-quality, analysis of current cyber events.

6.3.1 Active and Passive Situational Awareness for Cyber

A simple decomposition of SA modalities includes "active" and "passive," as shown in Table 6.2.

As shown in Table 6.2, an example use of M&S assisting cyber is automating response (Raulerson et al. 2014) to events that are beyond an operator's sensory or temporal capabilities. For example, while modern computer networks and subsequent cyberattacks grow more complex each year, analyzing associated network information can be difficult and time consuming. Network defenders, routinely unable to orient themselves quickly enough to determine system impact, might be helped by automated systems to find and execute event responses to minimize damage. Current automated response systems are mostly limited to scripted responses based on data from a single source.

6.3.2 Cyber System Monitoring and Example Approaches

Leveraging what we learned from Chapter 2's assessments, these evaluations provide a ready baseline for developing the metrics and measures for situation awareness. Most of the risk assessment frameworks provide enough information for this first step in developing SA. For example, DHS CSET (Department of Homeland Security [DHS]) or NIST's Baldridge Cybersecurity Excellence Builder process, in providing a cyber risk self-assessment, provides users with a baseline for their current use of cybersecurity policies and frameworks, helping an organization to shore up its resilience strategy before moving on to the technical monitoring. One approach (Amina 2012) performs adaptive cybersecurity analytics that include a computer-implemented method to report on network activity. A score, based on network activity, and using a scoring model, indicates the likelihood of a security violation.

Table 6.2 Situational Awareness and available M&S tools for improving cyber defense decision making.

Situational Awareness (SA) modality	M&S decision support improvement tools
Passive (i.e. "collect and collate") approaches	Understand network state via event ranking[a] and subsequent parsing (e.g. event management)
	Visual Analytics (VA) is a way of handling and making sense of massive data sets (e.g. Big Data)
Active approaches	Use of "spoofing" tools (e.g. mirror networks) to distract attackers and monitor their behavior
	Use of "Bots" to automatically respond, when authorized, to identified threats (Zetter 2014)

[a] Experimental results indicated that when administrators are only concerned with high-level attacks, impact assessments could eliminate a mean 51.2% of irrelevant data. When only concerned with high- and medium-level attacks, a mean of 34.0% of the data was irrelevant. This represents a significant reduction in the information administrators must process (Raulerson et al. 2014).

An additional SA approach is provided by Siemens (Skare 2013), in providing an integrated command and control user interface to display security-related data (e.g. log files, access control lists, etc.). In addition, this approach provides a system security level and interfaces with a user to update system security settings for an industrial control system based on the security-related data collected. This includes remote updating of access controls.

While there are many more examples of patented approaches (e.g. in Chapter 12) for helping with cyber evaluation, we will keep our focus to more specific M&S tools; leveraging conceptual models (e.g. OODA) is one way to keep this focus.

6.4 Cyber COAs and Decision Types

If the threat is unknown, or in a large system, and the impact of a certain COA is uncertain, M&S contributes by looking at the impact of an attack and determining mitigation strategies from there. As shown in Table 6.1, COAs leverage Table 6.1's metrics in determining the right level of model abstraction, and their associated metric(s). While Data Farming (Section 6.2.2.1) helps once the model and scenario are determined, it is still a challenge to decide which scenario best represents the future set of decisions, and associated COAs, for the projected cyber threat. Figure 6.8 provides an example "what, how, and why" diagram of the key cyber terrain,

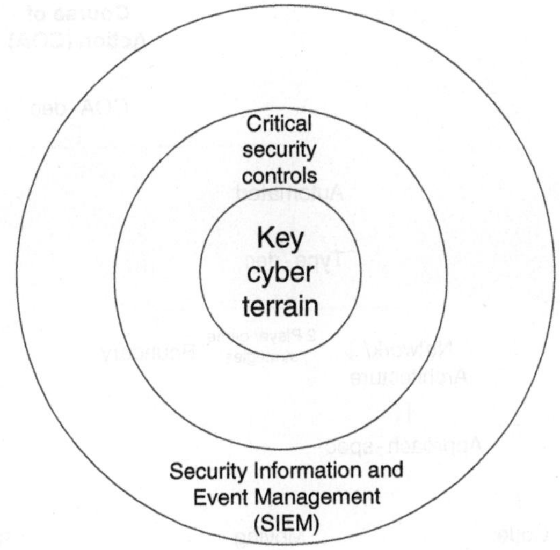

Figure 6.8 SIEM data, CSCs, and key cyber terrain – the what, how, and why of cyber decision making.

CSCs and SIEM data acquisition, for making decisions about protecting a cyber enterprise.

The dynamic nature of cyberattacks, their evolution in both frequency and effectiveness, requires a corresponding flexibility in security policy and associated technical responses to ensure real-time effectiveness. While a Disaster Recovery/Continuity of Operations (DR/COOP) plan is a key part of an organization's security planning, shorter-term attacks require a persistent SA and rapid response capability; automating some courses of action. An overview of a COA structure is shown in Figure 6.9.

As shown in Figure 6.9, automated and human-assisted COA implementations provide a simple demarcation between (i) the traditional human-assisted system (e.g. SIEM, etc.) and (ii) automated systems, which are called out by the Critical Security Controls (CSCs) and described by tools.

6.5 Conclusions

An immediate application of COA understanding is the building of training simulators, leveraging legacy military training knowledge, and inserting cyber into computer-assisted exercises (CAX) to determine COA effectiveness. Constructive simulation plays a role in developing CAX emulators and determining the high-risk scenarios where training is required. In addition, current events (e.g. Estonia, Georgia, etc.) provide examples of threat scenarios where modeling and training are required for effective future response.

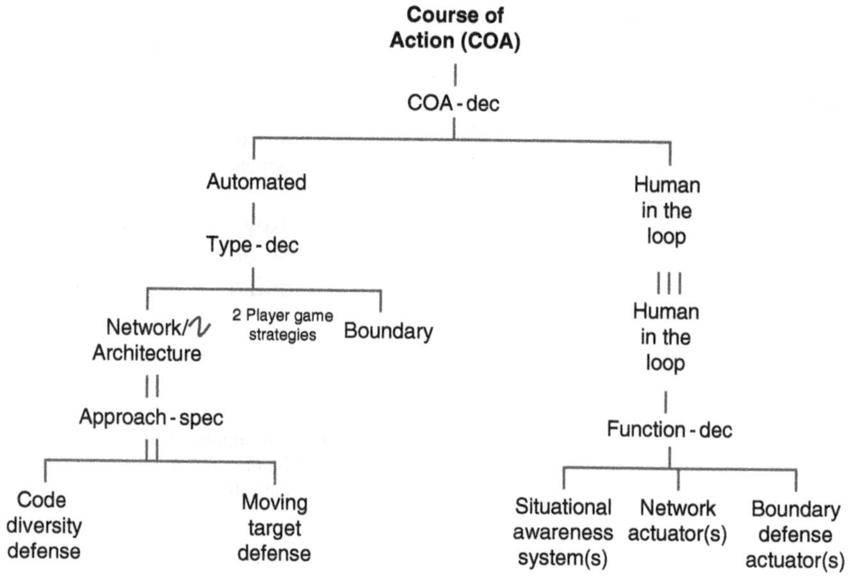

Figure 6.9 Course of Action implementations (automated and human assisted).

6.6 Further Considerations

Due to the broadly scoped cyber threat, formulating a defense can be a challenge. One approach to coordinating the multiple phenomenologies is to use a Network Tasking Order (Compton et al. 2010). The NTO, using known cyber metrics (Table 6.1) for DSS design, provides a measure for better understanding how secure we are over a particular threat scenario.

6.7 Questions

1 What is the difference between a COA for cyber and a COA for physical security defense planning?

2 What is the difference, in potential tools, between offline and on-line tools in doing an evaluation for EBCOTE?

3 What are the two main elements of AMICA cyber evaluation?

4 How does an attack graph inform a COA?

5 Where do process models fit into Figure 6.5's cyber DSS?

6 What are the key differences between passive and active SA?

7 How is a mirrored domain fundamentally different from a honey pot?

7

Cyber Computer-Assisted Exercise (CAX) and Situational Awareness (SA) via Cyber M&S

While automated COA responses are a goal for "simple" cyber operations, strategic planning and coordinated response still rely on human decision making. Detecting, reacting, and reconstituting cyber systems are therefore a function of human skill; skills that can be improved through training. These training requirements are translated by instructional system designers and subject matter experts to define the readiness competencies using Bloom's Taxonomy (Figure 7.1).

Developed by Benjamin Bloom (1994), Figure 7.1's Taxonomy divides educational objectives into three "domains": cognitive, affective, and psychomotor (sometimes loosely described as "knowing/head," "feeling/heart," and "doing/hands," respectively). Within the domains, learning at the higher levels is dependent on having attained the prerequisite knowledge and skills at lower levels. This parallels our training pipeline approach where "doing/hands" objectives are developed in foundational training; "feeling/heart" objectives are developed in sub-element validation and certification activities; and readiness training is achieved through "knowing/head" objectives.

One approach to "livening up" training is to add game-theoretic processes, modeled with moves and effects inspired by cyber conflict but without modeling the underlying processes of cyberattack and defense (Manshaei et al. 2013; Cho and Gao 2016). In addition, it is often pointed out that accurate predictions require good models of not just the physical and control systems, but also of human decision making; one approach being to specifically model the decisions of a cyber–physical intruder who is attacking the system and the system operator who is defending it – demonstrating the model's usefulness for design (Backhaus et al. 2013).

A goal of computer-assisted exercises (CAX) is to ensure that both individuals and teams are ready for stressing defense situations. One example, Crossed Shields (NATO Cooperative Cyber Defense Center of Excellence), an exercise from the NATO Cooperative Defense Cyber Center of Excellence (CDCCOE), uses cyber computer-assisted exercise (CAX) for cyber defense training; the

An Introduction to Cyber Modeling and Simulation, First Edition. Jerry M. Couretas.

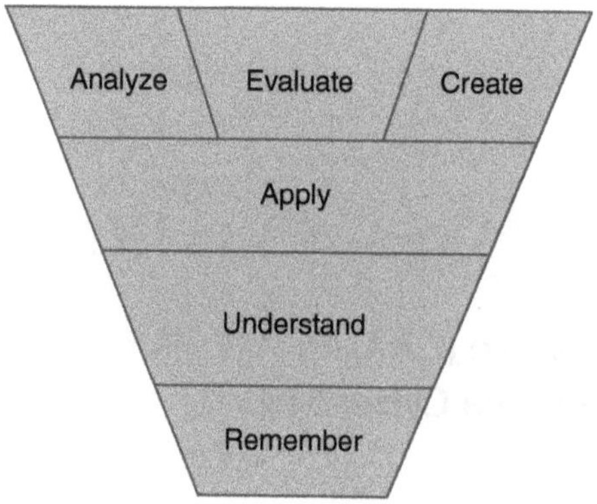

Figure 7.1 Bloom Taxonomy for learning domains.

goal of which is to teach personnel how to choose the most appropriate Course of Action (COA) in the event of an attack. In this chapter we make a distinction between training types for:

- Cyber for Cyber (C4C) – experts on cyber defense. Sometimes C4C can include former Information Assurance (IA) personnel evaluating IT system security.
- Cyber for Others (C4O) – operators tasked with dealing with the denied/degraded environments resulting from cyberattacks.
  - To date, the representation of cyber has been rare – experiments (e.g. COATS) introducing cyber into traditional command-level training simulation have been performed, but automated approaches are still developing – exercises seem to be the principle means of evaluating system security.

Using the C4C and C4O definitions to guide our look at cyber M&S for training, we make a distinction between the two different types of CAX:

- (C4C) "Cyber CAX" that focuses on Cyber Defense at a systems level (Scalable Network Defense Trainer[1] (NDTrainer [Chapter 12]).
- (C4O) Traditional CAX with cyber injections
  - Cyber Operational Architecture Training System (COATS) (Wells and Bryan 2015).
  - Cyber Operations Battlefield Web Services (CobWEBS) (Marshall et al. 2015).

In addition, we will discuss how these two types of CAX leverage the physical, informational, and cognitive elements[2] of Information Operations (IO) (Joint Chiefs of Staff 2014) to provide a basis for measuring cyber situational

1 http://web.scalable-networks.com/network-defense-trainer
2 Physical, Informational, and Cognitive.

awareness (SA) via cyber M&S; and how both kinds of CAX are commonly facilitated by M&S. We will also look at the different training Tiers (e.g. Global through individual) and available tools and metrics used to judge their performance.

As will be discussed in Section 7.2, SA is key to both being aware that a system is under cyberattack and taking defensive measures to protect the system. While the three layers of IO provide an initial reference for describing cyber SA (Robinson and Cybenko 2012), we can also leverage decades of Observe/Orient/Decide/Act (OODA) development, targeted for training pilots across the spectrum of air operations.

7.1 Training Type and Current Cyber Capabilities

In this chapter, we review traditional CAX, look at cyber injects into traditional CAX, evaluate cyber CAX, and look to understand combined traditional and cyber CAX. Table 7.1 provides a few examples of cyber training systems across the training Tiers; each training system referenced in Chapter 12.

Table 7.1's Traditional CAX and Cyber injections into Traditional CAX have the goal of reusing existing simulation platforms for operator (C4O) training. Cyber CAX and Training games, however, are relatively new. The Cyber CAX

Table 7.1 Group training capabilities.[a]

CAX type	Training system	Individual	Team	Regional	Global
Cyber Injections into Traditional CAX	COATS, CobWEBS		X	X	X
Cyber CAX	NDTrainer – (Exata/Antycip)		X	X	
	TOFINO SCADA Sim	X	X	X	
	CYNRTS (Camber)	X	X		
	HYNESIM (Diatteam)	X	X		
	CyberShield (Elbit)	X	X		
Training Games	NETWARS/Cyber City (SANS)		X	X	
	CyberCIEGE		X	X	
	MAST	X			

[a] References for each of the tools are provided in the Appendix.

and Training games have a goal of training pure cyber (C4C) personnel on systems that either are, or represent, actual devices of interest. Chapter 8 will talk about device emulation and its current evolution into simulation. Cyber CAX and training games are the current realization of this idea, providing users with scenarios so that they can experiment with possible courses of action (COAs). Chapter 6 covered COA evaluation, which is performed with Table 7.1's training system examples.

Each of Table 7.1's training systems, in providing a tool for COA evaluation, maintains the singular goal of increasing operator SA; or the ability to discern that their system is under cyberattack, and, once identified, determine an appropriate COA. This remains an unsolved problem in the commercial world. For example, in 2013, the median number of days attackers present on commercial victim networks, before being discovered, was 229 days, down from 243 days in 2012 (Mandiant 2014); indicating a persistently low SA.

7.2 Situational Awareness (SA) Background and Measures

With over 200 days to discovery of the average APT threat, an initial learning objective for cyber CAX is tactical situational assessment, the process of understanding one's environment that leads to SA and understanding. In addition, while there are two distinct audiences for cyber training (C4C, C4O), each group will require the development of SA to do their job. SA training, a goal of pilot training over the last several decades, provides an exemplar for cyber SA training. For example, in the course of air combat training, the US Air Force developed the Observe Orient Decide Act (OODA) loop (Boyd), with the observe–orient commonly ascribed to being SA development (Table 7.2).

As shown in Table 7.2, SA and understanding occurs over different time horizons, strategic and tactical, with different learning processes and objective outcomes. With SA formally defined as "The perception of the elements in the environment within a volume of time and space, the comprehension of their meaning, and the projection of their status in the near future" (Endsley 1995),

Table 7.2 Situational awareness learning – tactical and strategic processes and outcomes.

		Phase	
		Process	**Outcome**
Learning objective	Strategic	Sense making	Understanding
	Tactical	Situational assessment	Situational awareness

a few example SA frameworks for further metric development are listed in Chapter 12. In measuring both operator aptitude before/after cyber CAX training, SA measures provide an evaluation framework for the different training alternatives.

Due to adversarial success, there is clearly a need for training simulators that will be used to ensure cyber security. While cyber modifications are currently being developed to leverage existing simulators, it can be a challenge to retrofit CAX designed for command and control (C2) of conventional operations with cyber requirements. For example, a common approach to cyber training, currently, is to use a legacy trainer, and turn off the communications to simulate a denial of service (DOS) attack. Simply leveraging a higher-level simulation and turning off the communications misses the point of cyber, where attackers are more likely to minimize detection of their presence on a network and modify the integrity of data in order to shape operations (i.e. similar to IO). This also brings up the challenge of clearly defining metrics for the cyber domain; an area where policy (e.g. implementation of cyber security controls – Chapter 12) is often viewed as a solution from the management perspective. Training to ensure policy implementation will likely need to be unpacked to ensure clear communication across a cyber defense organization. In addition, this more nuanced use of cyber will likely require a tailored training simulator to meet this need and explore the possibilities.

While the challenges of cyber training are still being defined, we can now take a look at how operational exercises have been used in the air domain to confront similar SA and subsequent OODA development capabilities. Fortunately, the air domain has already tackled many of the structural training issues that cyber currently faces.

7.3 Operational Cyber Domain and Training Considerations

Cyber is at an early stage and still premature in clearly specifying the models that govern the dynamics of the cyber domain for CAX training. Figure 7.2 (Stine 2012) provides a notional interaction between the network, cyberspace, and mission operations that help inform both cyber injections into current CAX and the development of standalone cyber CAX.

Figure 7.2 provides just one view of an "as is" architecture, parts of which are emulated by the CAX in training future cyber defenders. The objective here is to emulate a real-world system, like that shown in Figure 7.2, with the right mix of Live–Virtual–Constructive (LVC) assets.

As shown in Figure 7.3, each of the LVC modes has different associated skill acquisition goals. We will see how the respective virtual and constructive injections are implemented in current CAX environments in evaluating other

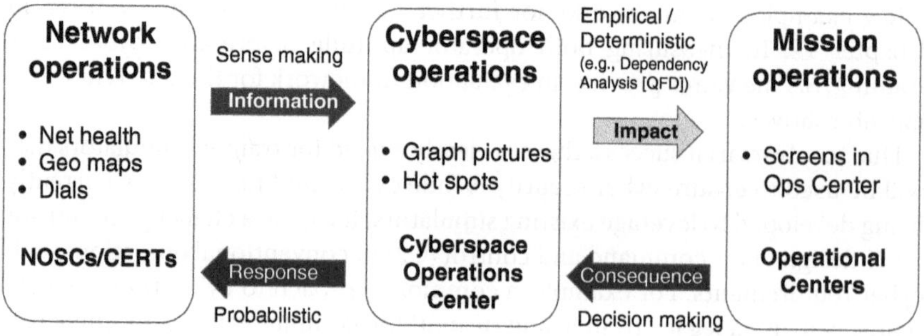

Figure 7.2 Network, cyberspace, and mission operations – information flows and events (Stine 2012).

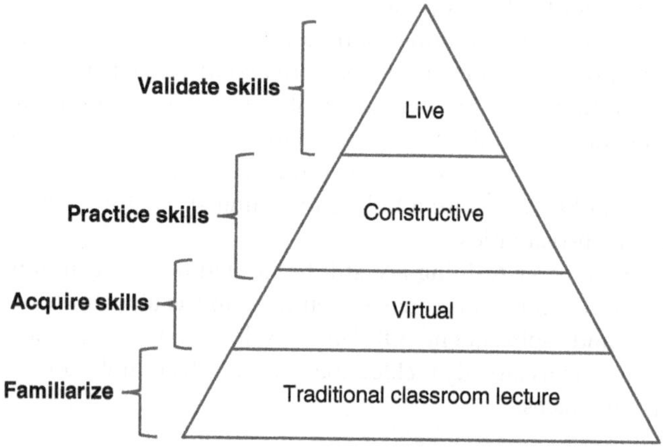

Figure 7.3 Live–Virtual–Constructive (LVC) and skills development.

Table 7.3 LVC contributions to cyber CAX realism.

LVC element	Description
Live	Injecting effects from operators into the simulation
Virtual	Injecting effects from ranges into the training simulation (e.g. COATS)
Constructive	Use of simulators to inject cyber effects into tactical exercise

domains (e.g. air training). Achieving the realism common to LVC for the air domain will require either producing simulations of realistic fidelity (constructive modeling) or consistently providing live injects (live/virtual modeling) into the CAX. A few considerations for achieving realistic fidelity and timeliness in building cyber M&S for defense CAX training are shown in Table 7.3.

Ideally, Table 7.3's techniques will contribute to cyber defense training simulators to keep pace with the fast-moving nature of the cyber domain. M&S, therefore, can be used to quickly put into operation rules that leverage recent field understanding of current cyber threats (Chapter 12) and provide these threats, for training, to our cyber defenders. The goal is then to provide realistic measures of SA improvement that will be used to inform both training updates and the future acquisition of material solutions to help with cyber defense.

7.4 Cyber Combined Arms Exercise (CAX) Environment Architecture

We are fortunate to have a baseline, in both LVC for bringing operator/range effects into our cyber simulations, and a functioning simulation architecture, via air training,[3] that leverages SA evaluation that is of interest for our cyber team. AMSP-03's (NATO 2014) Distributed CAX Environment Architecture (Figure 7.4) provides a CAX environment architecture as a generic construct, where teams may represent nations, corporations, or any other members of the training audience.

Figure 7.4 shows the overall CAX environment architecture and the interactions between the training audience, and the respective command levels (Table 7.4).

In addition to Table 7.4's description of the CAX components, the arrow color convention in Figure 7.4 provides the following:

- The red arrows represent the information exchange between the M&S tools. The different types of information should be considered for standardization (Chapter 5).

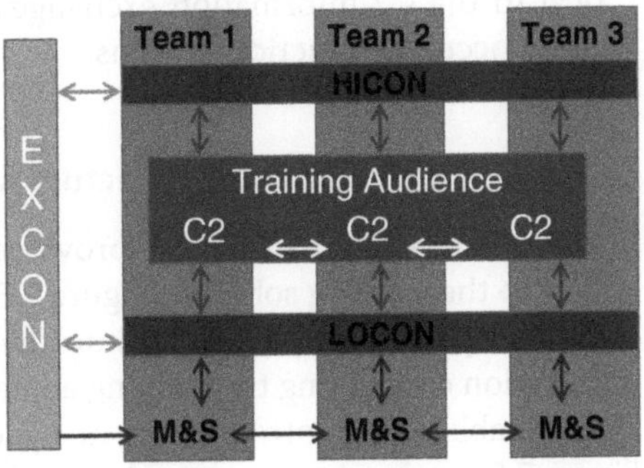

Figure 7.4 Generic CAX environment architecture.

3 Air training is currently mature enough to be provided in turn-key packages, with an airman entering a training program and a pilot coming out of the other end with guaranteed timeliness and quality attributes (http://www.lockheedmartin.com/us/products/mfts.html).

Table 7.4 CAX environment components.

CAX environment component	Description
Training Audience	The Training Audience plays staff in operation and performs C2
LOCON (Low Control Cell)	The LOCON (Low Control Cell) plays the subordinate command or units of the training audience. The LOCON provides reports based on the orders created by the training audience. The response Cell staff uses M&S to facilitate report generation that leverages data used for the system state data
Opposition Forces (OPFOR)	The OPFOR plays the opposing forces
HICON (High Control Cell)	HICON (High Control Cell) plays the training Tiers above the training audience level. The HICON provides directives and could request situational reports from the training audience
EXCON (Exercise Control Cell)	The EXCON (Exercise Control Cell) performs scenario execution and injects events/incidents planned in the MEL/MIL (Main Event List/ Main Incident List) in coordination with LOCON, the OPFOR, and the HICON.

- The brown arrows represent the information exchange between the EXCON staff and the role players. These information exchanges concern document exchange and control activities by any collaborative mode (Email, chat, phone, etc.).
- The blue arrows represent the information exchange between C2. These information exchanges are defined by the C2 community.
- The green arrow represents the information exchange between the Trainees and the LOCON and HICON. These information exchanges should be identical to the C2 information exchanges. Nevertheless, some simplifications could occur for practical reasons.

7.4.1 CAX Environment Architecture with Cyber Layer

Adding a cyber layer to a CAX provides the trainer control in adding cyber effects to the training solution (Figure 7.5).

By incorporating Cyber M&S into the generic CAX architecture, all communication concerning the training audience will be passed through the cyber layer, enabling the introduction of cyber effects in training. As shown in Figure 7.5, a cyber injection would target the green or the blue arrows by

- Confidentiality – Intended or unintended disclosure or leakage of information.
- Integrity – Creating false information.
- Availability – Degrading the flow of information, i.e. slowing down or preventing it.

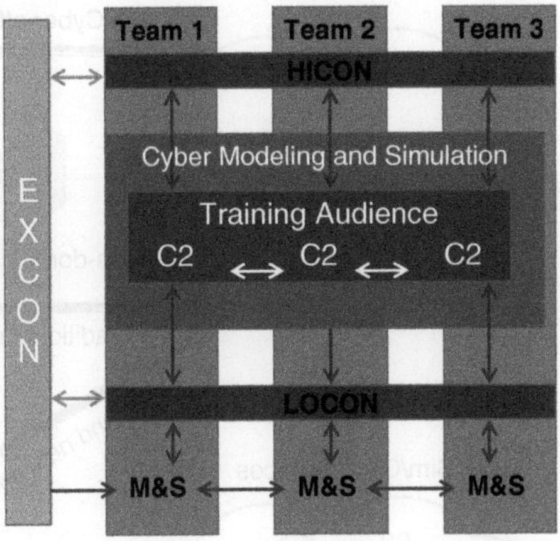

Figure 7.5 Generic CAX environment architecture including a cyber layer.

Table 7.5 Description of Red Computer Network Attack (CNA) on Blue systems to demonstrate degradation effects on operator workstations.

Red cyber engagements (in cyber range) against Blue entity causes user detectable effects (e.g. Blue Screen of Death (BSoD), CPU memory utilization, etc.)	
Desired effects: Blue operator workstation degradation	**Generated effect**: BSoD and CPU memory utilization
Challenge: Propagating effect through exercise infrastructure	**Attack direction**: Red on Blue
Demonstration audience: Chief Information Security Officer (CISO) and staff	**LVC category**: Cyber (Live/Virtual) to Cyber (Virtual)
Target network: Live asset in the cyber environment, generic end-user workstations	**Effect network**: Private LAN – generic blue end-user workstations

The first type of cyber threat, leaking system information, might be injected either by HICON or LOCON as well as by a communication simulation system. Similarly, the latter two might be achieved by the use of communication simulation systems, which are able to model different types of communication disturbances, affecting data integrity or reducing system availability (Table 7.5).

As shown in Table 7.5, COATS and the Network Effects Emulation System (NE2S) (Morse et al. 2014a, b) provide cyber effects to a training audience. In addition, COATS provides a distributed simulation architecture in providing cyber effect injects, from cyber ranges, as shown in Figure 7.6.

While Figure 7.6 shows how COATS provides range-based cyber injects into training simulations, it is also possible to use constructive modeling as a proxy for the range-provided effects. This opens up the opportunity to control the level of cyber effects in a command-level simulation. For example, the NE2S

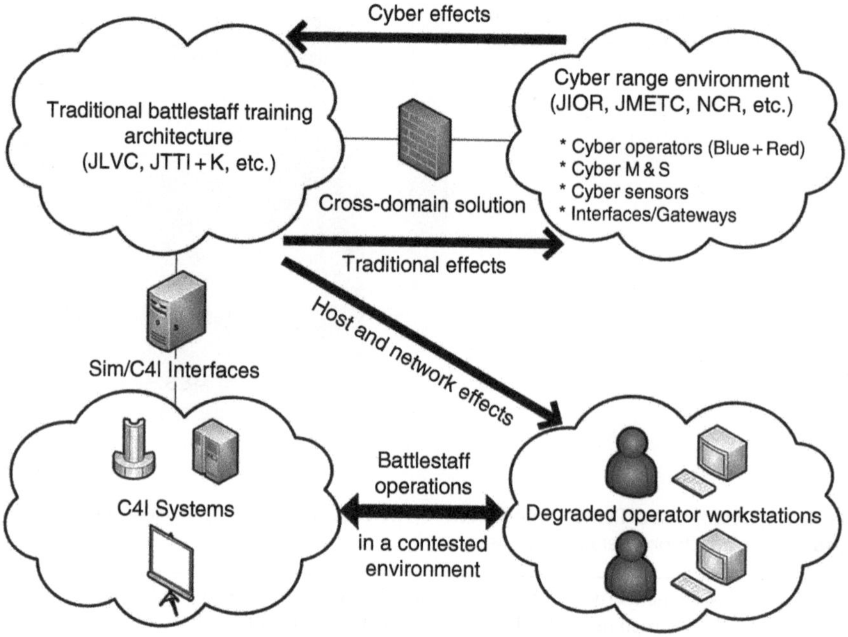

Figure 7.6 COATS cyber injection architecture.

(Bucher, 2012) is used to add cyber effects to a command-level simulation. This approach is currently targeted at higher-level (e.g. Global and Regional) exercise simulations.

7.4.2 Cyber Injections into Traditional CAX – Leveraging Constructive Simulation

The traditional CAX is targeted to strategic, operational, and/or tactical-level training audiences. In this type of exercise, cyber injections are just one among several threads that the audience has to deal with. Because of the novelty, and lack of real-world controllability, cyber is often handled via white cards, at present, where a cyberattack can be scheduled for a particular effect, in the course of an exercise focused on a broader attack model. The aim of including a cyber injection in a traditional CAX is

- Training of the audience's strategic, operational and/or tactical skills.
- Raising the audience's awareness of cyber threats to enhance the ability to recognize and mitigate Cyber threats, which might be done through virtual/ constructive injects of cyber effects (i.e. range effects).

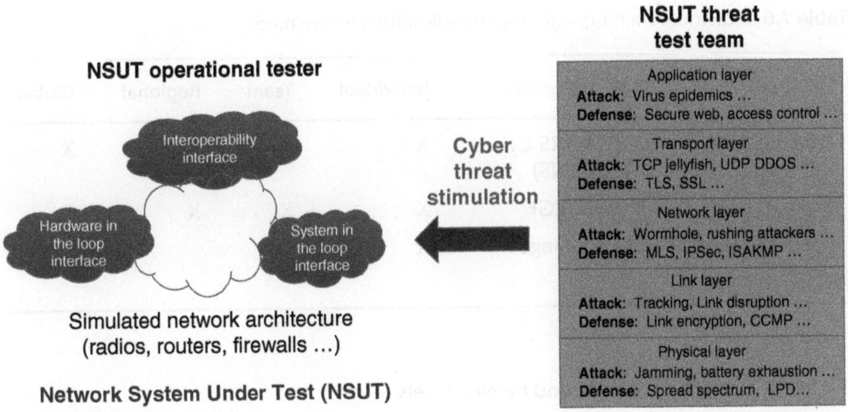

Figure 7.7 Network emulation (StealthNet) injection into Network System Under Test (NSUT) (Bucher 2012).

StealthNet (Torres 2015), developed by Exata for the Test Resource Management Center (TRMC), has the goal of using five tiers of the OSI hierarchy level as constructive simulation layers for exercising architectures on a cyber range (Figure 7.7).

Figure 7.7's example (Bucher 2012), from the test and evaluation (T&E) community, is also viable as a constructive simulation input for CAX. Some examples of the effects for different simulation injects include:

- Loss of communication nodes and lines of communication (e.g. DOS).
- Loss of fidelity of sources of communication.
- Partial loss of information.
- False or compromised information (i.e. key differentiator for cyber CAX).
- Restrictions in bandwidth.

The aspects of NATO and Multinational CAX are listed and explained in AMSP-03,[4] but these are also valid for CAX on a smaller (e.g. national or organizational) level.

7.4.3 Cyber CAX – Individual and Group Training

We use the term Cyber CAX for CAX that is intended to train the audience in the application of methods of cyber defense. Training focuses on exercising technical personnel, both military and non-military subject matter experts for cyber defense.

4 AMSP-03: M&S standard profile for NATO and Multinational Computer-Assisted eXercises with Distributed Simulation.

Table 7.6 Individual training – games and administrator training.

CAX type	Training system	Individual	Team	Regional	Global
Training Games	NETWARS Cyber City (SANS)	X	X	X	X
	CyberCIEGE	X	X	X	X
	MAST (Singh)	X			

Table 7.7 Example Cyber CAX and training levels.

CAX type	Training system	Individual	Team	Regional	Global
Cyber CAX	NDTrainer – (Exata/ Antycip)	X	X	X	
	TOFINO SCADA Sim	X	X	X	X
	CYNRTS (Camber)	X	X		
	HYNESIM (Diatteam)	X	X		
	CyberShield (Elbit)	X	X		

Individual training approaches include CyberCIEGE (Thompson and Irvine 2011), where video gaming technology is used to train candidate cyber defenders. SANS NETWARS (SANS) is a more common gaming platform used in the cyber training community (Table 7.6).

More traditional CAX span from individual to large group training, providing the opportunity to evaluate teams at multiple levels (Table 7.7).

Table 7.7's Cyber CAX and training levels provide the different tools on the market today for training individuals to large organizations. In addition, Table 7.7's focus on Cyber CAX provides an overview on how current operators, and cyber professionals, are currently training for improving their SA of the cyber terrain.

7.5 Conclusions

Cyber SA, one of the differentiating elements of the cyber domain, has its best definition in Information Operations (IO) (e.g. physical, informational, and cognitive), at present. Quantitative approaches to cyber SA still rely heavily on computer science (e.g. graph theory, etc.) for network description; cyber SA, as a human process, is still a work in progress. Similarly, CAX for cyber remains a developing domain at the time of this writing.

7.6 Future Work

With the goal of Cyber CAX being cyber operator SA, overall scenario and exercise design might consider using fusion levels (Waltz 2000) to evaluate the operator's ability to assess his/her situation; and how this assessment ability overlaps with operator performance. Examples measures include:

- Operator (C4C or C4O) estimated attack surface(s).
- Actual attack surfaces, as measured by network evaluation tools.

Finding common ground between objectively/technically measurable phenomena (e.g. fusion levels – not involving human judgment) and the somewhat qualitative metrics from human training (e.g. SA metrics) will provide trainers with a few different tools to measure C4C/C4O operators in evaluating their performance. Assessing human vs. machine performance will also provide an objective evaluation of what can be automated. For example, Guo and Sprague (2016) show a Bayesian replication of human operator's situation assessment and decision making performing very well in a simulated area reconnaissance war game.

Along with using objectively quantifiable data to evaluate cyber defenders, our future Cyber CAX should leverage standard threats when building both exercise scenarios and evaluation COAs. In addition, a future Cyber CAX, leveraging standard threats and scenarios, might look something like a combination of Morse et al. (2014a, b) and Bucher's (2012) cyber training architectures, distributed to capture best-of-breed capabilities wherever they reside on the net, and expanded to all of the LVC dimensions, so that each of the training levels can be provided through one training architecture.

7.7 Questions

1. Which of the Bloom Taxonomy's educational objectives are most important for cyber training and why?

2. Why are the IO domains a useful reference measure for cyber computer-aided exercises?

3. How should an organization match its CAX training and SA development goals?

4 Looking at the LVC skills development pyramid (Figure 7.3), where should the strongest emphasis be placed for building SA skills?

5 Why is it a good idea to combine C4C and C4O training, via a COATS-like approach?
A What are the challenges for this kind of training?

6 In developing a cyber CAX, what are the additional LVC considerations to make the training more realistic?

8
Cyber Model-Based Evaluation Background

Evaluating cyber systems is usually a trade between the realism of a live experiment and the speed provided by a representative model-based simulation; broadly described in terms of scale, scope, and fidelity. Characterizing these cyber systems, to achieve fidelity and validity through physical models of the system of interest, is challenged by limitations in the flexibility and scalability of a physical model. Abstracting on these physical systems, usually in software (Guruprasad et al. 2005), results in a flexible environment to construct computer networks (Rimondini 2007).

Emulation, due to fidelity and known validity, is often how operational network testing is currently practiced. Simulation, using a constructive representation of the system, has scalability and flexibility benefits. In deciding the merits of emulation vs. simulation for a particular evaluation, the system evaluator should consider the fidelity, scalability, and flexibility (i.e. scope) tradeoffs required for the test object's modeling scenarios (Table 8.1).

Table 8.1 provides example descriptions for each type of system parameterization. Fidelity is best provided with the system under test, next best is an associated emulation. While emulators provide scalability beyond what physical systems offer, models provide both the flexibility and scalability characteristic of current computer-based systems.

8.1 Emulators, Simulators, and Verification/Validation for Cyber System Description

Cyber simulation is currently practiced on "cyber ranges;" (Davis and Magrath 2013) computational clean rooms used to prevent viruses, worms, and other logical malefactors from infecting surrounding systems. Leveraging these ranges for cyber security evaluation is usually performed via emulators, replicating the operational configuration to be protected in a realistic scenario. This 1:1 relation, emulator to real-world, is ripe for using M&S to provide the n:1

An Introduction to Cyber Modeling and Simulation, First Edition. Jerry M. Couretas.

Table 8.1 System attributes – flexibility, scalability, and fidelity.

Attribute	Description
Flexibility (i.e. scope)	Ability to reconfigure environment – this might be evaluating the model of interest for another use case and associated validity evaluation.
Scalability	Scalability has to do with altering the size of the network of interest. While "scalability" is a factor for virtualizing cyber-range environments, scalability is also the value proposition provided by many of the contemporary "cloud services" for remote computing, a potential advantage for model-based simulations.
Fidelity	The most accurate fidelity is provided by the real system, hence the use of "system in the loop" for many of the most critical testing applications. Abstracting on the real system will necessarily reduce fidelity, affecting usability of the associated model, based on its intended use.

extensibility that software based modeling can provide. This section will therefore look at example emulators, simulators, and their potential combination; along with a quick look at corresponding verification and validation (V&V) process that provides the operational community with more confidence that cyber models represent the real world.

8.2 Modeling Background

Emulators, or recreations of systems to mimic the behavior of the original system, are used for hardware and software systems where the original is inaccessible due to cost, availability, security, or obsolescence issues. While emulators are commonly used in current cyber M&S, simulators provide potential cost savings for improving the scale and scope of system models. Increasing scale will help us simulate the actual number of entities in a system; increasing scope will capture the requisite variety (e.g. components, system states, etc.) that makes modeling real system attack surfaces such a challenge.

The basic modeling relation, *validity*, refers to the relation between a model, a system, and an Experimental Frame (EF) (Zeigler and Nutaro 2016). Validity is often thought of as the degree to which a model faithfully represents its system counterpart. However, it is more practical to require that the model faithfully captures system behavior only to the extent demanded by the simulation study's objectives. The concept of validity answers the question of whether it is impossible to distinguish the model and system in the EF of interest; the behavior of the model and system agree within acceptable tolerance.

One of the key benefits of emulators is validity, with respect to the system of interest. Implied in duplicating a system are predictive and structural validity. In predictive validity, the emulator provides not only replicative validity, but also the ability to predict as yet unseen system behavior. Ideally, this is a state-for-state duplication of the system of interest. To do this, the emulator, and subsequent model, needs to be set in a state corresponding to that of the reference system.

The term accuracy is often used in place of validity. Another term, fidelity, is often used for a combination of both validity and detail. Thus, a high fidelity model may refer to a model that is both high in detail and in validity. However, when used this way there may be a tacit assumption that high detail alone is needed for high fidelity, as if validity is a necessary consequence of high detail. In fact, it is possible to have a very detailed model that is nevertheless very much in error, simply because some of the highly resolved components function in a different manner than their real system counterparts.

8.2.1 Cyber Simulators

Maintaining model to system validity, at the same level as current emulators, requires verification of the subsequent model/description. EFs are one way to capture the intended uses that make the emulator/model a successful description in the domain of interest. Table 8.2 provides both the conceptual

Table 8.2 Conceptual definitions of activities and modeling and simulation framework (MSF) equivalents.

Activity	Description	M&S formalization
Verification	Process to determine if an implemented model is consistent with its specification	Simulation correctness, a relation between models and simulations, uses the verification process for proving simulator correctness to generate model behavior. This approach certifies simulator correctness for any model of the associated class.
Validation	Process of evaluating model behavior using known use cases	There is a relation, called "validity in a frame," between models and real systems within an EF. Validation is the process of establishing that the model behaviors and real system agree in the frame in question. The frame can capture the intended objectives (extended to intended uses), applicability domain, and accuracy requirements.
Abstraction	Detail reduction process to replicate only what is needed in a model	Abstraction is the process of constructing a lumped model from a base model, intended to be valid, for the real system in a given experimental frame.

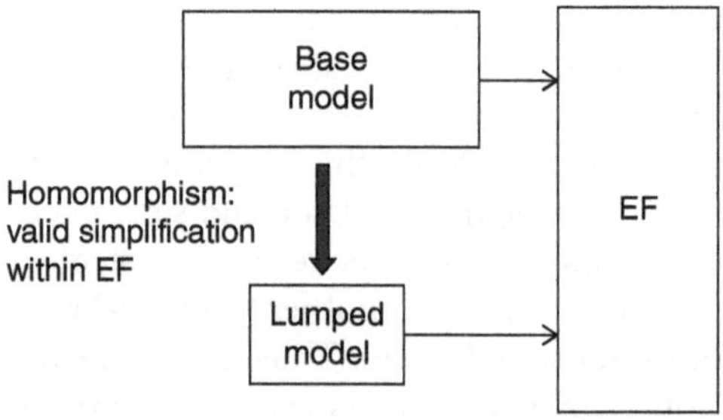

Figure 8.1 Validity of base and lumped models in Experimental Frame (EF).

definitions of verification/validation/abstraction and their Modeling and Simulation Framework (MSF) formalizations (Zeigler and Nutaro 2016).

Besides validity, as a fundamental modeling relationship, there are other relations that are important for understanding modeling and simulation work. These relations have to do with EF use in model development. Successful modeling can be seen as valid simplification. We need to simplify, or reduce the complexity, of cyber behaviors if scalable descriptions, mimicking the large-scale systems that currently support our daily lives, are to be developed for planning and analysis. But the simplified model must also be valid, at some level, and within some EF of interest.

As shown in Figure 8.1, there is always a pair of models involved – call them the base and lumped models (Zeigler et al. 2000). Here, the base model is typically "more capable" and requires more resources for interpretation than the lumped model. By the term "more capable," this means that the base model is valid within a larger set of EFs (with respect to a real system) than the lumped model. However, the important point is that within a particular frame of interest the lumped model might be just as valid as the base model. Figure 8.1's morphism approach provides a method to judge base and lumped model equivalence with respect to an EF.

Abstracting from base to lumped model presents a challenge in cyber system description due to the necessary context (e.g. representative Course of Action) that the cyber system portrays. In addition, the moving parts represented by the attack cycle, or representative taxonomy (e.g. ATT&CK, MACE, etc.) might be considered at this stage, concerning the role that vulnerability evaluation will play in what the overall model describes. Subsequent metrics might also be considered during the lumping to ensure that the final representation provides the analytical insight that the user is focused on.

While keeping the systems' considerations in mind in abstracting the model, a key enabler for developing more capable simulators that build on Figure 8.1's morphism example is the ability to construct lumped models, from individual base models, that represent more abstract system of systems (SoS) (Figure 8.2) (Zeigler and Nutaro 2016).

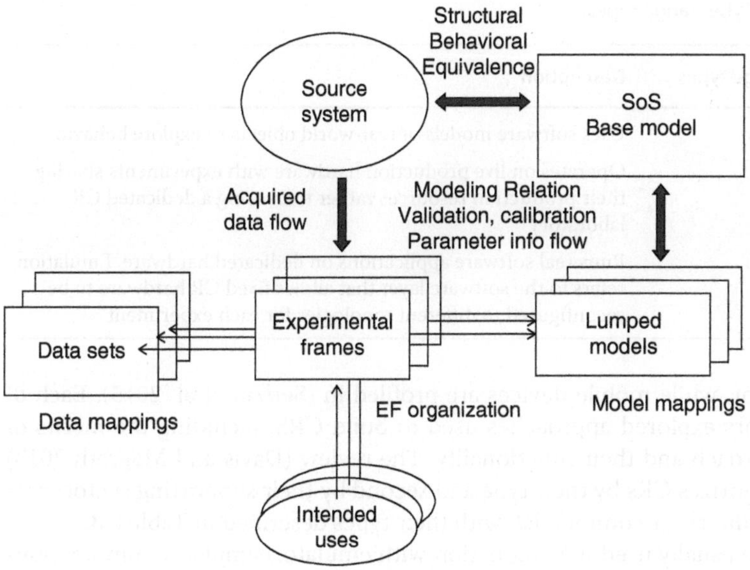

Figure 8.2 Architecture for System of Systems (SoS) Verification and Validation (V&V) based on M&S framework.

Figure 8.2 shows how the data sets, intended uses, and EFs roll up to provide both SoS base and lumped models; an aspirational approach for current cyber M&S. While Figure 8.2 provides an overview of all of the components for a constructive simulation system that incorporates mappings required for emulator incorporation into simulators, test beds are still the primary means for testing emulator/simulator combinations for cyber–physical systems. In addition, Kim et al. (2008) developed the DEVS/NS-2 Environment for network-based discrete event simulation within the DEVS framework for easy capture of EF and other DEVS-based concepts presented here: NS-2 sometimes viewed as a parallel to an emulation with its topology and node/link configuration requirements.

8.2.2 Cyber Emulators

Cyber ranges (CRs) often include elements of both emulation and simulation of networks. For example, exercises usually include simulation of an attack on the network and its progress, and countermeasures by human defenders. In addition, CRs leverage this paradigm, which is often used to evaluate a system or technology concept.

Several CRs are provided by Davis and Magrath (2013). Cyber VAN (Chadha et al. 2016) provides an example of a portable range for standard computing

Table 8.3 Cyber range types.

Cyber range types	Description
Simulation	Uses software models of real-world objects to explore behavior
Overlay	Operates on live production hardware with experiments sharing their production resources rather than using a dedicated CR laboratory
Emulation	Runs real software applications on dedicated hardware. Emulation refers to the software layer that allows fixed CR hardware to be reconfigured to different topologies for each experiment.

equipment, while mobile devices are profiled in (Serban et al. 2015). Each of the authors explored approaches used to build CRS, including the merits of each approach and their functionality. The review (Davis and Magrath 2013) first categorizes CRs by their type and second by their supporting sector: academic, military, or commercial, with their types described in Table 8.3.

CRs are usually used in conjunction with emulator/simulator combinations for evaluating overall systems.

8.2.3 Emulator/Simulator Combinations for Cyber Systems

While the theory is in place for providing valid cyber M&S, in practice, cyber-physical systems are still evaluated as independent functional manifestations on CRs composed of virtual machines. Modeling provides for the evaluation of system states, in the abstract, apart from the strict function calls found in software system regression testing. Leveraging both system state understanding and attack graphs (Jajodia et al. 2015), modeling provides the opportunity to clearly identify the system states that lead to the vulnerabilities enumerated by Cam (2015), and discussed more clearly in the MITRE ATT&CK framework. In addition, mapping a system's state space also provides for the application of quality control approaches (e.g. factorial design, etc.) to better enumerate a system's state combinations and potential vulnerabilities.

Attacks against cyber (IT) and physical (e.g. industrial control system, actuators, etc.) systems are usually evaluated independently, at present, emulating the individual systems of interest and their respective logical anomalies. The SoS that makes up a Cyber–Physical System (CPS) is therefore only evaluated component by component, intended uses of which may not account for system states found in combination.

Thus, there is a pressing need to evaluate both cyber and physical systems together for a rapidly growing number of applications using simulation and emulation in a realistic environment, which brings realistic attacks against the defensive capabilities of CPS. Without support from appropriate tools and

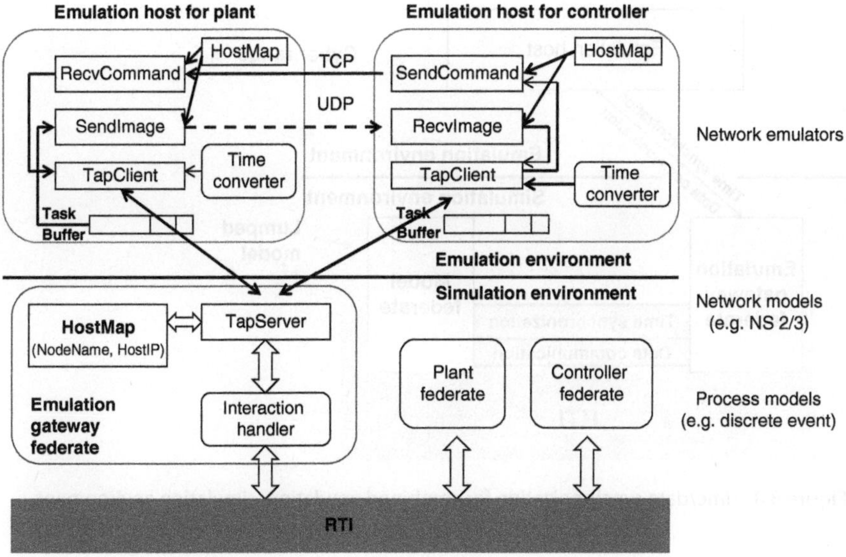

Figure 8.3 Emulator–simulator combination for Cyber–Physical System.

run-time environments, this assessment process can be extremely time-consuming and error-prone, if possible at all. Integrating simulation and emulation together, in a single platform for security experimentation, exists at the concept stage,[1,2] further proving out the need for such an environment and bringing out some of the considerations required for a full-scale application. Major components for a mixed simulation/emulation environment include (Yan et al. 2012):

1) Modeling environment for system specification and experiment configuration.
2) Run-time environment that supports experiment execution.

At run time, the cyber simulator/emulator provides time synchronization and data communication, coordinating the execution of the security experiment across simulation and emulation platforms (Figure 8.3). As previously discussed (Chapter 7), COATS provides this hybrid simulation/emulation combination via communicating cyber effects from an emulation test bed to a more traditional command-level training simulation environment.

An extension of individual system testing, similar to the COATS (Morse et al. 2014a, b) example for incorporating cyber effects into training, requiring

1 http://seer.deterlab.net/
2 http://isi.deterlab.net/

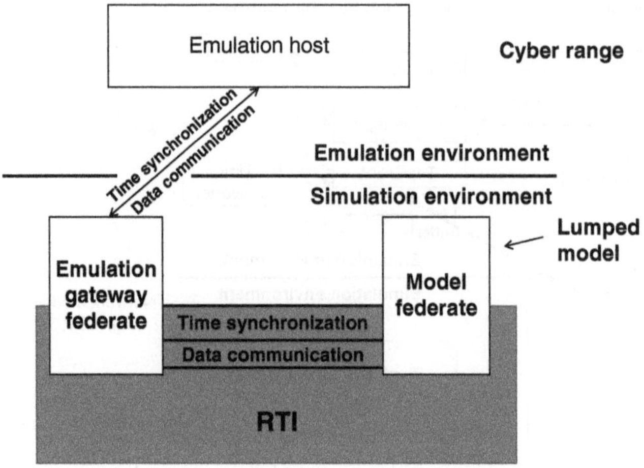

Figure 8.4 Time/data synchronization for combined emulation–simulation environment.

a combination of CRs (i.e. emulated environments) and trainees, there is also interest (often for engineering purposes) to combine range emulators with constructive simulations. Simulations have the potential to provide "cheap" scalability not available in other "real" applications.

A key addition in Figure 8.3's combined emulation and simulation environment is the synchronization of temporal and data communications. As shown in Figure 8.4, this was a gap in DETERLAB and is handled in iSEE (Yan et al. 2012) setup (Figure 8.4).

While Figure 8.4 provides a method for combining emulation and simulation environments for generalized cyber–physical testing, a cost and portability target for this work is to increasingly leverage structured modeling of both individual and SoS', for valid evaluation of associated cyber systems.

8.2.4 Verification, Validation, and Accreditation (VV&A)

Leveraging simulations, and using EFs in particular, shows that the user has a clear picture of acceptance criteria for the final M&S system (Roza et al. 2010). The *acceptance goal* is to convincingly show that an M&S system will satisfy its purpose in use. This abstract acceptance goal is translated into a set of necessary and sufficient concrete acceptability criteria; criteria for which convincing evidence is obtained. General Methodology for Verification and Validation (GM-VV) defines three classes of acceptability criteria for M&S artifacts, called VV&A *properties* (Figure 8.7) that each address and provide a set of assessment metrics for a specific part of an M&S artifact (Roza et al. 2013) (Table 8.4).

Table 8.4 Verification, Validation and Accreditation (VV&A) properties.

Acceptability criteria for M&S artifacts	Verification, Validation, and Accreditation (VV&A) properties
Utility	Properties used to assess the effectiveness, efficiency, suitability, and availability of an M&S artifact in solving a problem statement in the problem world. Utility properties address aspects such as value, risk, and cost.
Validity	Properties that are used to assess the level of agreement of the M&S system replication of the real-world systems it tries to represent, e.g. fidelity. Validity properties are also used to assess the consequences of fidelity discrepancies on the M&S system utility.
Correctness	Properties that assess whether the M&S system implementation that conform to the imposed requirements, and is free of error and of sufficient precision. Correctness metrics are also used to assess the consequences of implementation discrepancies on both validity and utility.

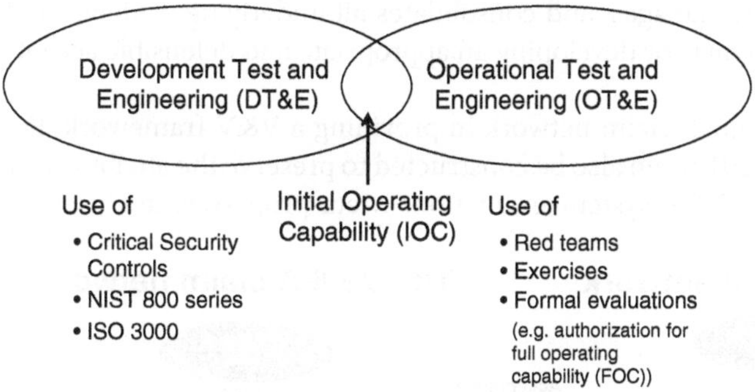

Figure 8.5 Development vs. operational testing – verification and validation.

While the GM-VV (Roza et al. 2013) provides an overall process for system evaluation, each of the respective steps requires a tool to clarify whether the tool is verified (meets the specifications/requirements you have written – "Did I build what I said I would?") and valid (addressed the business needs that caused you to write those requirements – "Did I build what I need?"). One way of representing system development incorporating these considerations is shown in Figure 8.5.

One of GM-VV's tools to accomplish V&V is the goal claim network (Figure 8.6).

The VV&A goal–claim network is an information and argumentation structure rooted in both goal-oriented requirements engineering and claim–argument–evidence safety engineering principles. The left part of the goal–claim network is used to derive the acceptability criteria from the acceptance goal; and design solutions for collecting evidence to demonstrate that the M&S system, intermediate product, or result satisfies these criteria. The acceptance goal reflects the VV&A needs and scope (e.g. system of interest, intended use). Evidence solutions include the specification of tests/experiments, referent for the simuland (e.g. expected results, observed real data), methods for comparing and evaluating the test/experimental results against the referent. Collectively, they specify the design of the V&V EF used to assess the M&S system and its results. When implemented, the EF produces the actual V&V results. After a quality assessment (e.g. for errors, reliability, and strength), these results can be used as the items of evidence in the right part of the goal–claim network. These items of evidence support the arguments that underpin the acceptability claims. An acceptability claim states whether a related acceptability criterion has been met or not. Acceptability claims provide the arguments for assessing whether or to what extent the M&S system and its results are acceptable for the intended use. This assessment results in an acceptance claim inside the VV&A goal–claim network.

Ideally, the goal–network is built in a top-down manner and the claim network in a bottom-up manner, as indicated by the rectangular arrows in Figure 8.6. However, in practice, the total VV&A goal–claim network is built iteratively as indicated by the circular arrows. The VV&A goal–claim network as such encapsulates, manages, and consolidates all underlying evidence and argumentation necessary for developing an appropriate and defensible acceptance recommendation.

In addition to the goal–claim network in providing a V&V framework, the GM-VV (Roza et al. 2013) can also be constructed to preserve the attributes, or meta-properties, used for system evaluation. Meta-properties are used to

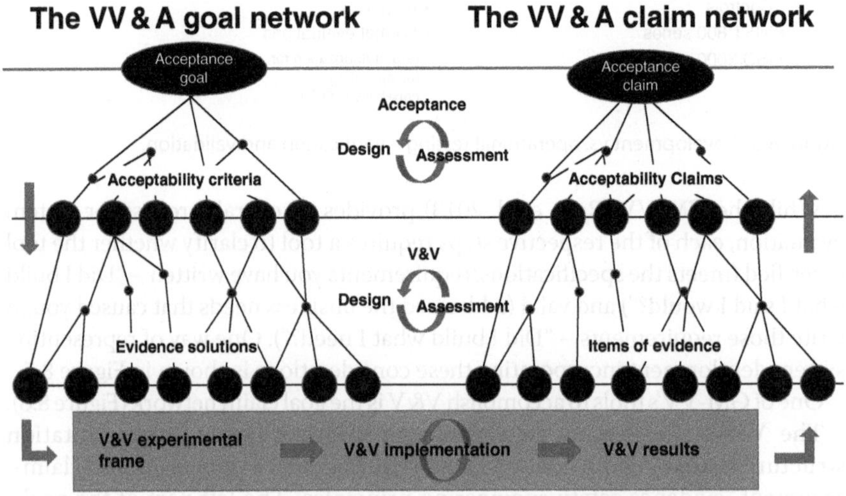

Figure 8.6 VV&A goal–claim network structure.

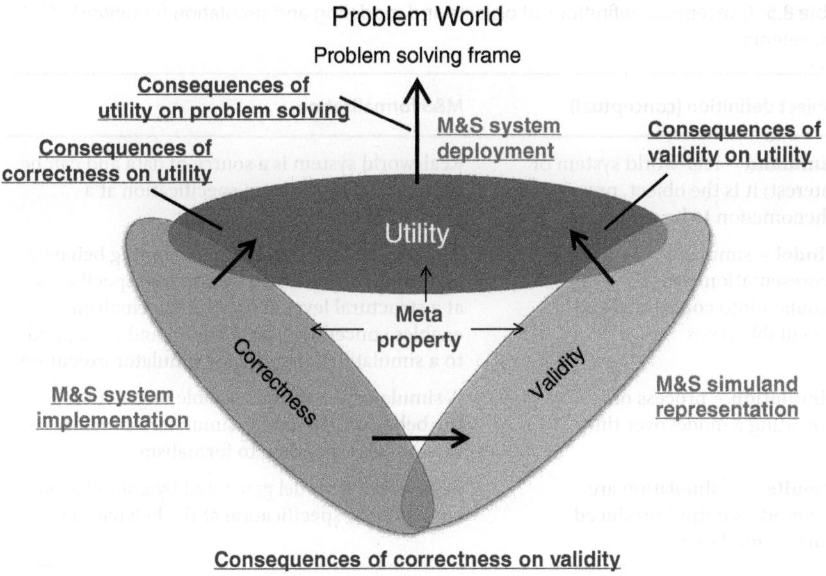

Figure 8.7 Utility, validity, correctness, and meta-properties relationship diagram.

assess the level of confidence with which the utility, validity, and correctness have been assessed, i.e. the convincing force of the evidence for these three properties. Meta-properties typically include aspects such as completeness, consistency, independence, uncertainty, and relevance (Figure 8.7).

Figure 8.7 expands on the type of analysis possible, once a standard, reusable foundation for simulation (e.g. DEVS formalism) is relied upon for developing the representative component and system models (Table 8.5).

Table 8.5 provides some basic M&S definitions extensible to describing cyber systems. Additional work on quantifying the uncertainty (Grange and Deiotte 2015) in representations holds promise for increasing developer understanding on "how good" a model-based simulation platform is for a particular cyber evaluation task.

8.3 Conclusions

The body of M&S theory exists for developing

1) EFs that represent COAs.
2) Mapping real-world, emulator-like, descriptions to mathematical objects (e.g. lumped models).
3) Verifying and Validating that a cyber simulation is correct within the context of a given EF.

Table 8.5 Conceptual definitions of objects and modeling and simulation framework (MSF) equivalents.

Object definition (conceptual)	M&S formalization
Simuland – real-world system of interest; it is the object, process, or phenomenon to be simulated	Real-world system is a source of data and can be represented by a system specification at a behavioral level
Model – simuland representation, broadly grouped into conceptual and executable types	A model is a set of rules for generating behavior and can be represented by a system specification at a structural level. A modeling formalism enables conceptual specification and is mapped to a simulation language for simulator execution
Simulation – process of executing a model over time	A simulator is a system capable of generating the behavior of a model; simulators come in classes corresponding to formalisms
Results – of simulation are the model's output produced during simulation	Behavior of a model generated by a simulation constitutes a specification at the behavior level

Providing valid simulations of cyber systems is still a manual process, ensuring that the respective EF describes the COA of interest, and making the argument that intended use has been met for V&V purposes. Fortunately, the GM-VV ties up the combination of EFs, component verification, and project management into an overall structure usable to construct any kind of simulation (including cyber). GM-VV is a Simulation Interoperability Standard's Organization (SISO) standard. Having gone through a multiyear vetting process, GM-VV is a good resource to leverage for verifying and validating cyber models.

8.4 Questions

1 Where do requirements fit into the constructing of a cyber model?

2 How is abstraction, from real world to model, usually accomplished?

3 What is the difference between a base and lumped model?

4 When is a cyber model verified?

5 What are the main issues in validating a cyber model?

6 What does an Experimental Frame (EF) provide the cyber modeler to facilitate V&V efforts?

7 Why is a goal–claim network a good approach for documenting model validity?

9

Cyber Modeling and Simulation and System Risk Analysis

"Cyber Risk," the other side of the enabling technologies provides time-saving conveniences that we have come to expect, has only recently added the word "risk." Much like the safety engineering movement that arose after we came to expect automobiles to go at unprecedented speeds and distance (e.g. seat belts only became mandated in the 1970s), cyber is now moving toward prescriptive policies to ensure positive user experience. This quality vs. security question is pertinent to cyber, much as it was to the auto industry a generation ago. A new wrinkle with cyber, however, is the active agency of the perpetrator compromising software vulnerability; failure rates proportional to attacker's skill.

9.1 Background on Cyber System Risk Analysis

Due to practitioners usually associating system "quality" with software quality (Ivers 2017), or verification of a system's design and subsequent measures of performance, it is sometimes actually easier to design a testing regimen based on known vulnerabilities and level of interest in hacking them. While this is one instance of system quality, I believe that the more comprehensive quality concept that developed around both the product and process development in the automotive industry (National Academy of Engineering 1995) is the kind of thinking necessary to develop secure cyber products.

For example, looking at Figure 9.1's Bathtub curve, both random and "wear out" failures are predictable for a given material with known characteristics. In addition, for consumer packaged goods manufacturing, the bathtub curve is a "rule of thumb" used for introducing any new product into the manufacturing process.

While Figure 9.1's simple depiction of product life is an attractive mental model for any system in production, there are a few extra considerations for cyber systems (Table 9.1).

An Introduction to Cyber Modeling and Simulation, First Edition. Jerry M. Couretas.

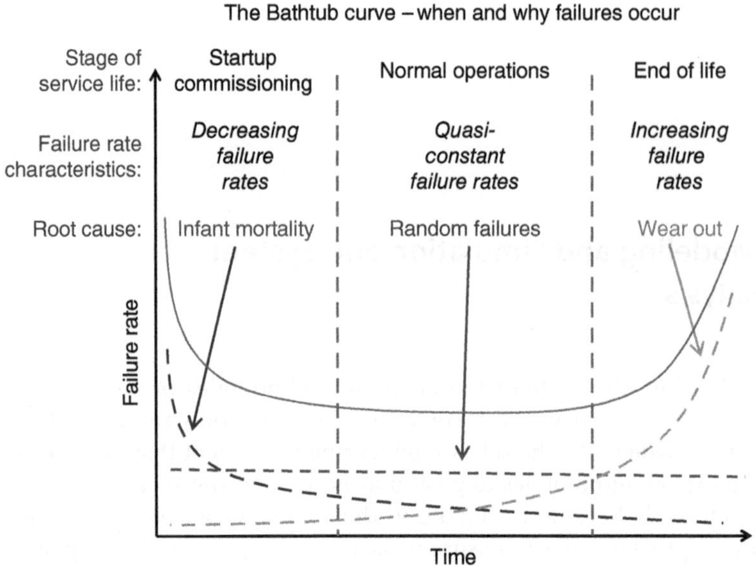

Figure 9.1 The bathtub curve – when and why of failures.

Table 9.1 Cyber modeling and the bathtub failure curve.

Enterprise components and bathtub curve for cyber	Description
People	Attacker/defender skill are independent variables in determining system failure rate; whether an exploit succeeds
Policy/Process	Patching processes may lead to "windows of vulnerability," adding discrete discontinuities to the constant failure rate section of the middle section of the bathtub curve
Process	Sampling is proportional to, and determinative of, knowledge acquisition on the part of both attackers and defenders; not a material depletion event stream
Technology	Lack of material/assembly failure that makes up much of the infant mortality section of the bathtub curve for manufacturing of consumer packaged goods, the foundation of the bathtub curve – cyber systems under inspection are assumed to work until they are decommissioned, or exploited

One of the key differences between manufactured product lifecycles and cyber systems is the active perpetrator. In addition, separating the "wear out" from the "random" failures, in a non-stationary environment, makes the situation much more challenging; requiring a baseline that essentially requires any kind of anomaly to be detected.

While a high-level example of an attacker is provided in Figure 3.7 (Chapter 3), taking it up a level, and consequent to improving cyber security, is including cyber security into project management's nexus of systems engineering and project planning (Kossiakoff et al. 2011), as shown in Figure 9.2.

As shown in Figure 9.2, systems engineering and project planning, done jointly, are required for preventing cyber anomalies in the first place. In addition, while the International Standards Organization published ISO 31000 to handle risk, the standard needs to be tailored for use in the cyber domain. This often includes looking at the overall system, its context, or reasons for particular construction, and how an attacker might look at the system. Fortunately, most systems have the common "V" construction (Figure 9.3), resulting in metrics being assigned at each of the respective construction phases.

As discussed in Table 9.3, there are multiple security metrics available for evaluating systems, depending on how much we know. The assumption here is that cyber system evaluation is a black box analysis, with little known about the system, where a general, time-based approach works as a first-order analysis.

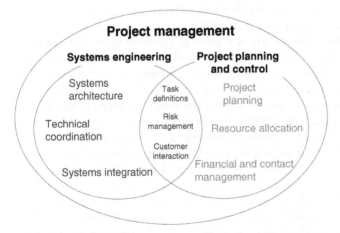

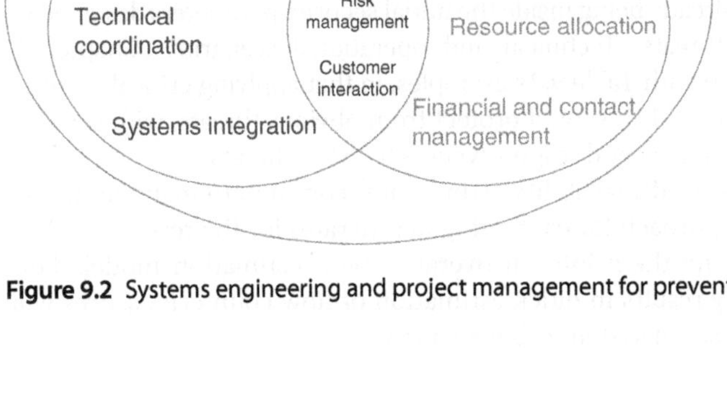

Figure 9.2 Systems engineering and project management for preventive cyber security.

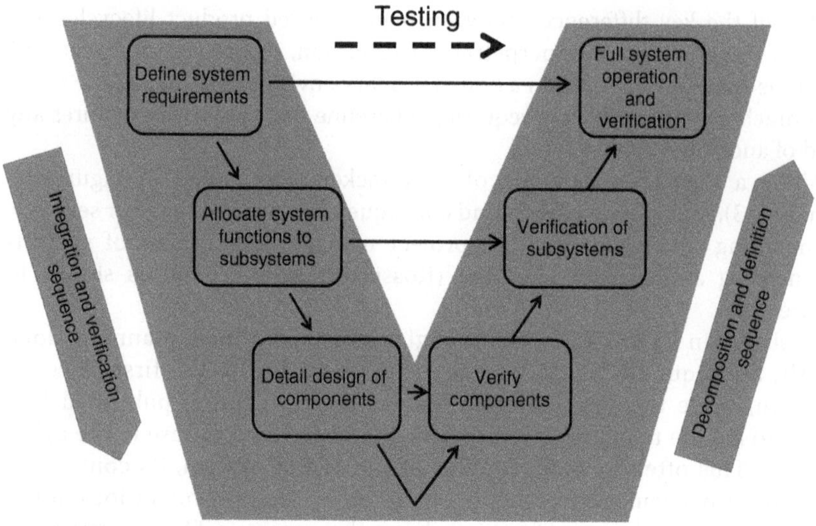

Figure 9.3 The "V" model of systems engineering.

9.2 Introduction to using Modeling and Simulation for System Risk Analysis with Cyber Effects

A methodology that evaluates pre-event cyber security risks, based on an enterprise's People, Processes, and Tools (PPT), is proposed here to proactively secure critical information that compose a company's technical and management differentiators. The "people" part of cyber vulnerability, well-known as the most common threat vector, is a challenge to model.

One example of risk evaluation is "war gaming" cyber, via M&S. This naturally brings to mind attackers/defenders. Cyber is novel in that it has at least one more dimension than standard games – team members, insiders, may be a source of risk. Some insider attack scenarios (Serdiouk 2007) are shown in Table 9.2.

Table 9.2's "people" threats may be some of the most challenging to handle, due to the perpetrators already being inside the initial security perimeters designed to keep out external threats. Technical and operational scenarios (Chapter 4, Table 4.3) will overlap with Table 9.2's examples, in that applying critical security controls (CSCs [Chapter 12]) should protect from obvious threats, with training to maintain both awareness and responsiveness to cyber threats.

Leveraging widely used availability estimations, common to manufacturing, we will provide an approach for estimating failure rates for the respective PPT domains and combine them into an overall exploit estimation model. This approach's flexibility results in quick estimation of how countermeasures will contribute to increases in system cyber security.

Table 9.2 Internal threat scenario examples.

Threat scenario	Description
1	Insiders who make "innocent" mistakes and cause accidental disclosures. Probably the most common source of breached privacy. For example, overheard hall conversations, misclassified data, email, etc.
2	Insiders who abuse their record access privileges. Individuals that have access to data and violate the trust associated with that access.
3	Insiders who knowingly access information for spite or for profit. When an attacker has authorization to some part of the system, but not to the desired data and gains access to that data by other means.
4	The unauthorized physical intruder. The attacker has physical entry to points of data access but has no authorization to access the desired data.
5	Vengeful employees and outsiders, such as vindictive patients or intruders, who mount attacks to access unauthorized information, damage systems, and disrupt operations.

9.3 General Business Enterprise Description Model

Evaluating a system's availability for mission performance, a solved problem for manufacturing quality control, is currently an issue in assessing IT and cyber physical system risk. Leveraging known systems engineering concepts, from data analysis to modeling mean time to failure (MTTF), provides cyber security designers with modeling and simulation (M&S) tools to describe and evaluate the people, process and technology (PPT) domains that compose a modern information-based enterprise. In addition, organizing these quality control techniques with entity-relational methods provides tools to coordinate data and parameterize enterprise evaluation models (Couretas 1998a, b). Similarly, Model-Based Systems Engineering (MBSE [Friedenthal et al. 2011]), an entity-relational approach for system description, is being used by the Department of Defense's (DoD) Defense Information Services Agency (DISA) to build the Joint Information Environment (JIE), or future cross-service network. Similarly, this approach applies to overall enterprise cyber risk evaluation (Couretas 2014).

DISA uses MBSE as a top-down approach to design the JIE. We propose a similar approach, in using an MBSE-like structure, to describe "as is" enterprises, modeling the respective PPT domains in terms of their "attack surfaces" (Manadhata and Wing 2008), with security metrics specified based on system/component vulnerabilities of interest (Table 9.3).

Table 9.3's metrics for core, attack graph, and temporal security evaluations give the system evaluator an "outer bound" of the system's current security

Table 9.3 Example security metrics.

Security metric	Examples	
Core	CVSS	Common Vulnerability Scoring System is an open standard for scoring IT security vulnerabilities. It was developed to provide organizations with a mechanism to measure vulnerabilities and prioritize their mitigation. There has been a widespread adoption of the CVSS scoring standard within the information technology community. For example the US Federal government uses the CVSS standard as the scoring engine for its National Vulnerability database (NVD[a])
	TVM	Total Vulnerability Measure is an aggregation metric that typically doesn't use any structure or dependency to quantify the security of the network. TVM is the aggregation of two other metrics called the Existing Vulnerabilities Measure (EVM) and the Aggregated Historical Vulnerability Measure (AHVM).
Probability (Attack-Graph)	AGP	In Attack Graph-based Probabilistic (AGP) each node/edge in the graph represents a vulnerability being exploited and is assigned a probability score. The score assigned represents the likelihood of an attacker exploiting the vulnerability given that all pre-requisite conditions are satisfied.
	BN	In Bayesian network (BN) based metrics the probabilities for the attack graph is updated based on new evidence and prior probabilities associated with the graph.
Structural (Attack-Graph)	SP	The Shortest Path (SP) (Ortalo et al. 1999) metric measures the shortest path for an attacker to reach an end goal. The attack graph is used to model the different paths and scenarios an attacker can navigate to reach the goal state which is the state where the security violation occurs.
	NP	The Number of Paths (NP) (Ortalo et al. 1999) metric measures the total number of paths an attacker can navigate through in an attack graph to reach the final goal, which is the desired state for an attacker.
	MPL	Mean of Path Lengths (MPL) metric (Li and Vaughn 2006) measures the arithmetic mean of the length of all paths an attacker can take from the initial goal to the final goal in an attack graph.
Time Based (General)	MTTR	Mean Time to Recovery (MTTR) (Jonsson and Olovsson 1997) measures the expected amount of time it takes to recover a system back to a safe state after being compromised by an attacker.
	MTTB	Mean Time to Breach (MTTB) (Jaquith 2007), represents the total amount of time spent by a red team to break into a system divided by the total number of breaches exploited during that time period.
	MTFF	Mean Time to First Failure (MTFF) (Sallhammar et al. 2006) corresponds to the time it takes for a system to enter into a compromised or failed state for the first time from the point the system was first initialized.

[a] NVD has a repository of over 45 000 known vulnerabilities and is updated on an ongoing basis (National Vulnerability Database 2014).

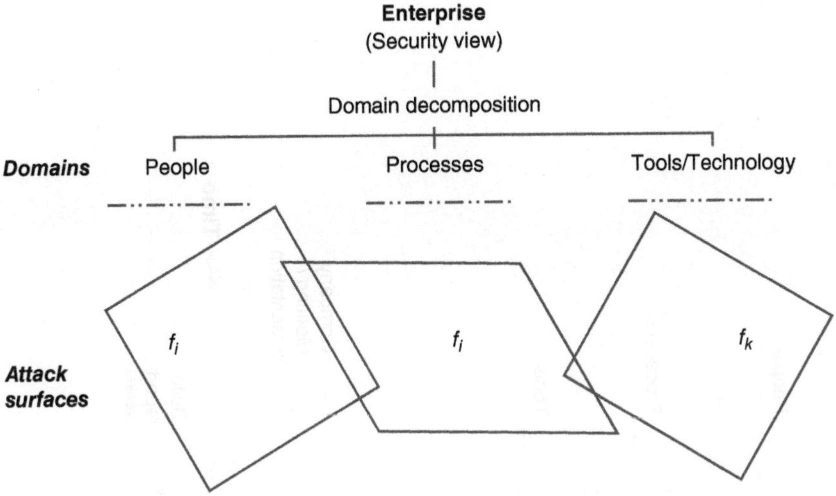

Figure 9.4 Enterprise security view – people, process, and technology domains.

state, providing a rough guide for tailoring the domain-based threat vectors for more specific evaluation(s). Using domains to narrow the scope for specific threat vectors results in the ability to estimate risk within individual domains. For example, estimating a "failure rate" for a sample population describes how long it will take for an exploit to occur (i.e. mean time to exploit [MTTE], t_{MTTE}), for either an individual domain (e.g. People[1]/Process/Technology) or, in combination, for the overall enterprise (Figure 9.4).

As shown in Figure 9.4, the security view for an enterprise decomposes into example PPT domains. Within each domain are the individual elements that compose the domain's attack surface; characterized by an example exploit probability, $\lambda_n \mathrm{d}t$, for a given slice of time, dt. In this case, each domain's attack surface is characterized by individual domain exploit rates, λ_n, and their combinations, $(f_i^* f_j)$.

9.3.1 Translate Data to Knowledge

The goal, as exemplified in Figure 9.5, is to evaluate/audit risk of an organization, and develop a report for leadership to understand where they stand, in terms of current cyber vulnerabilities.

1 On 21 May 2014, The Ponemon Institute released an Insider Threat study that showed 2/3 of respondents have trouble discerning what kind of activities constitutes a threat (http://www.ponemon.org/blog/ponemon-institute-and-raytheon-release-new-study-on-the-insider-threat).

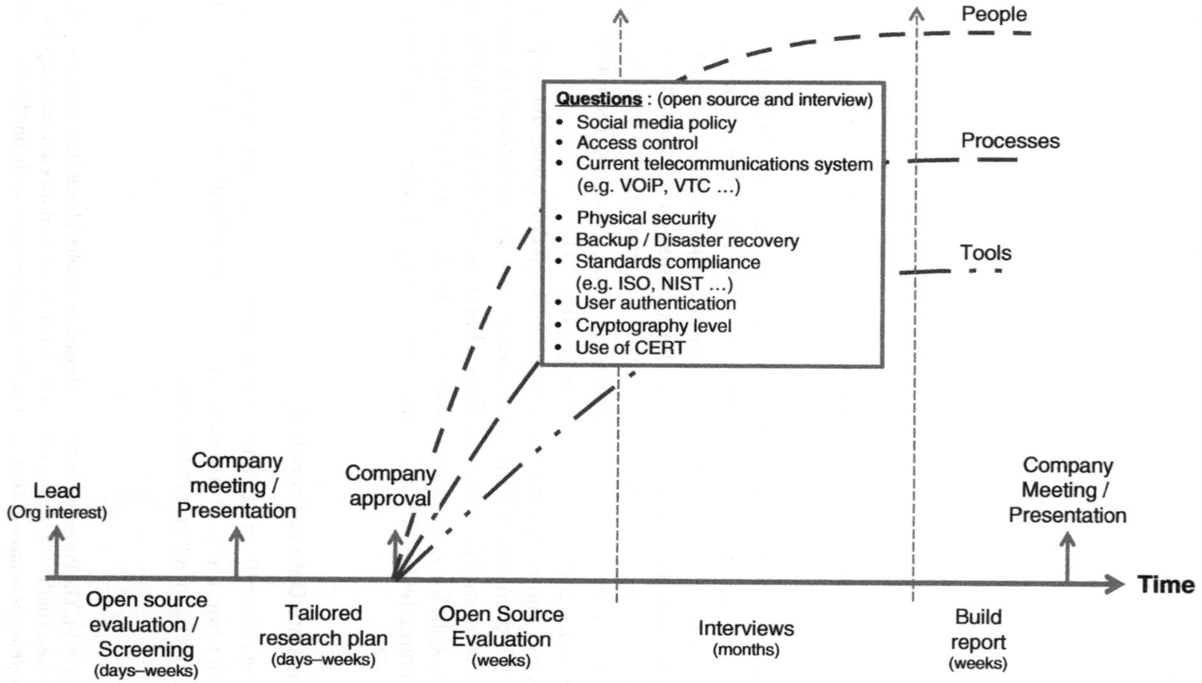

Figure 9.5 Audit timeline – people/process/technology evaluation.

Table 9.4 Enterprise evaluation – areas and time periods.

Evaluation areas	Collectables	Time periods
People	Anomaly detection	Weeks – months
Policy/Process	Performance changes	Months – quarters
Tool/Technology	Allowed on network (e.g., white listing)	Hours – days

As shown in Figure 9.5, people, policies, processes and tools that compose the enterprise are periodically evaluated via surveys and interviews (i.e. "collectibles") (Table 9.4).

In addition, Figure 9.5's surveyed areas are the data used to develop a description of the enterprise's interconnections and vulnerabilities (Figure 9.6).

While Figure 9.6 brings out the input/output description of the respective enterprise domains, we will use the system's entity-relationship model, called a System Entity Structure (SES), to build a taxonomy-like structure for describing the overall system. Using the people, processes, and technologies that make up the enterprise, use failure rate estimations for each domain, and in combination, to understand the current system's MTTE. Given the above enterprise security description, we use combinations of Policy/Training/Technology to construct alternative strategies. For example, gathering the data, as shown in Figure 9.9, is stored in a SES to provide an overall enterprise model to be exercised for risk/vulnerability of the system in question (Figure 9.7).

As shown, Figure 9.7 leverages the SES (Zeigler et al. 2000) as the enterprise "As Is" Graph Description to organize disparate People/Processes/Tools descriptions:

- Tree Structure example is currently an "AND" graph, where each of the decomposed entities has its own failure rate, that is used to contribute to the overall failure rate for each key node of the Enterprise (e.g. PPT).
- Decompositions can also include "OR" specialization nodes, where alternative people, process, or technology implementations are available.
- Graph Structure is formally called a System Entity Structure; used to describe an enterprise for evaluation (Couretas 1998a, b).

An example questionnaire, filled out to check the vulnerability of an enterprise network, is provided in Table 9.5.

Figure 9.8 provides an example of how the failure rates (e.g. λ) are rolled up for the respective PPT/technology vulnerabilities.

As shown in Figure 9.8, λ is the failure rate for each respective domain (e.g. people, processes, tools/technologies), or one of its components. Representing each of the people/process/tools with an exponential distribution results

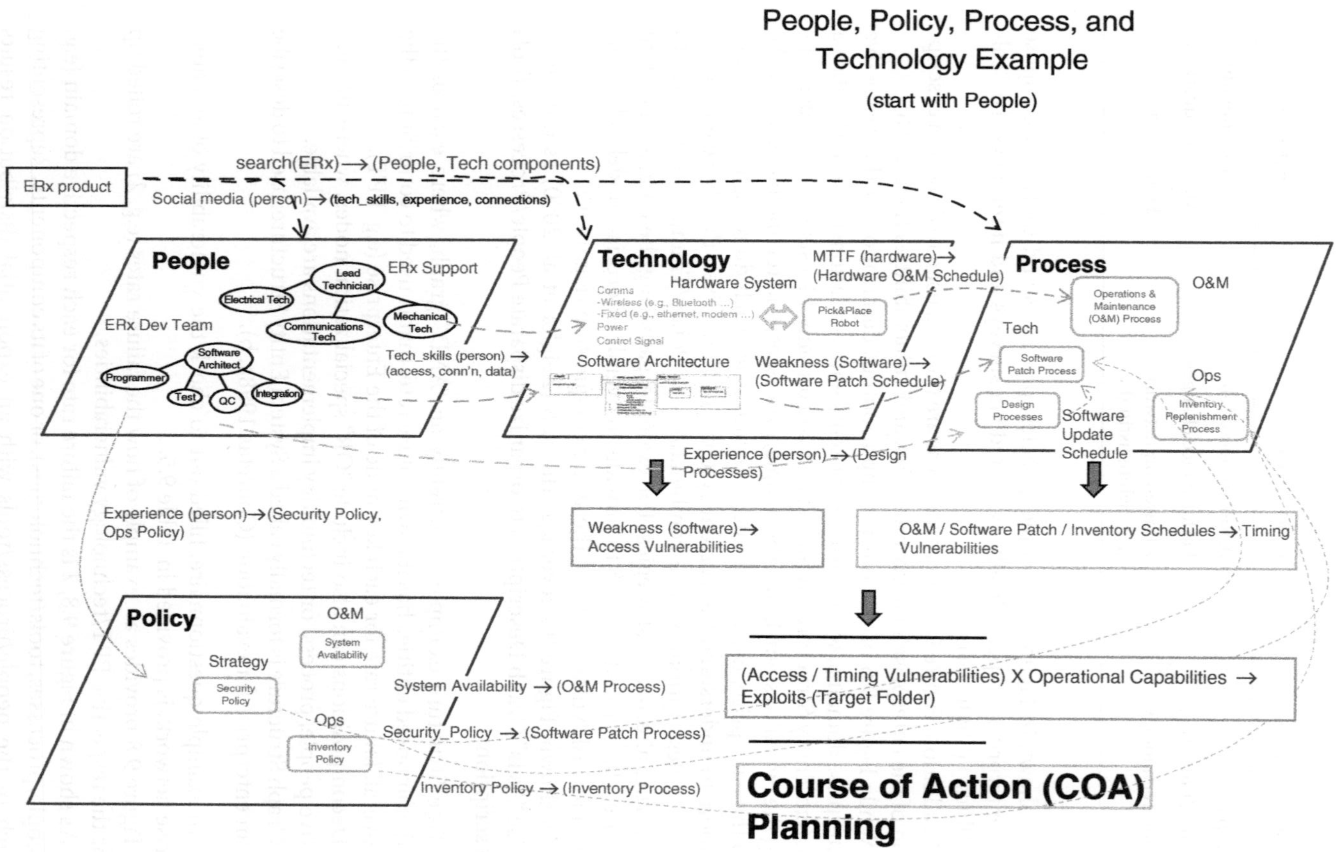

Figure 9.6 Enterprise connections (people/policy/process/technology) and COA planning.

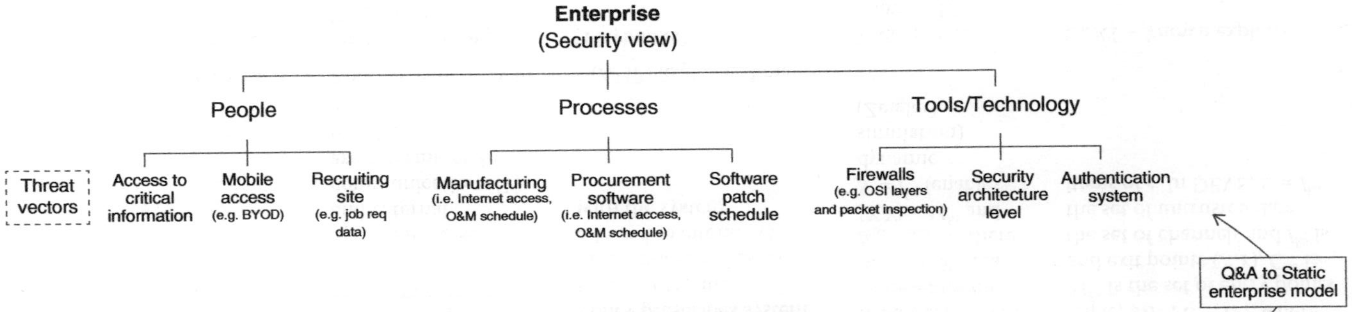

Use the Q&A process to develop an information structure amenable to modeling:

	People	Processes	Tools
Who	• System access		• User authentication
What	• Personally Identifiable Information (PII) • Social media	• Critical information • High volume (e.g. manufacturing)	
When	• System access	• Maintenance schedule • Patch schedule • Software updates	
Where	• Fixed site • Mobile		
Why	• Business system access • Technology system access		• Secure Sockets Layer (SSL)
How	• Recruiting • Screening		• Security architecture level • Firewall – monitoring and control

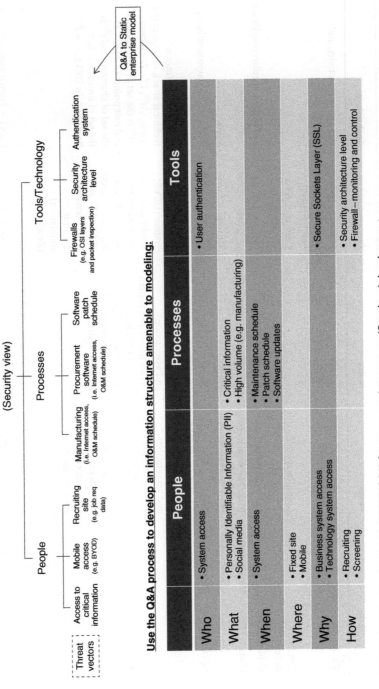

Figure 9.7 Question & answer process (Q&A) for system entity structure (Ontology) development.

Table 9.5 People/policy/process/technology example breakdown (vulnerability analysis).

		People	Policy	Process	Technology
Weaknesses		Access, Associations, and adherence	Locations, suppliers, and servicing	Software updates, service times, and maintenance	Access, connectivity, and storage
Crit Info Access		Login/password		Connection via servicing	
Vulnerability		Personal issues	Adherence issues		SANS 20 controls/compliance
Units	Enterprise	Risk, No. weaknesses, Vulnerabilities, Critical information access			
	Components	Entities, relations	Rules	Timing	Facts
Equations	Enterprise	Risk = 1 – Π*i*(1 – *ri*) | *i* = framework component, *r* = component risk with the risk for each component being – Risk = 1 – (weaknesses/(weaknesses + critical information access))			
	Components	People have access and associations, which are represented in an entity-relation diagram; static. System Accesses (*X*) and external communications (*Y*) are determined by user base	Policy constrains accesses and associations, Policy prescribes system behavior (*S*) and determines how users are allowed to interact (δ_{ext}) with the system	Discrete Event System (DEVS) Model for process representation: $M = \{X, Y, S, ta, \delta_{ext}, \delta_{int}, \lambda\}$ where $\{X,Y\} = M^{Es}$ and $S_i = I^{Es}$ (enables dynamic simulation) (Zeigler)	Manadhata defines a technical attack surface as the triple, $\{M^{Es}, C^{Es}, I^{es}\}$, where M^{Es} is the set of entry points and exit points (*X*,*Y*), C^{Es} is the set of channels and I^{Es} is the set of untrusted data items of s. In DEVS, $S_i = I^{Es}$
Tools	Enterprise	Analytic Hierarchy Process (AHP) (e.g. Checkmate)			
	Components	Link Analysis (e.g., Palantir, i2, etc.)	"Playbooks"	Visio, power point, JIRA	CERT – known exploits

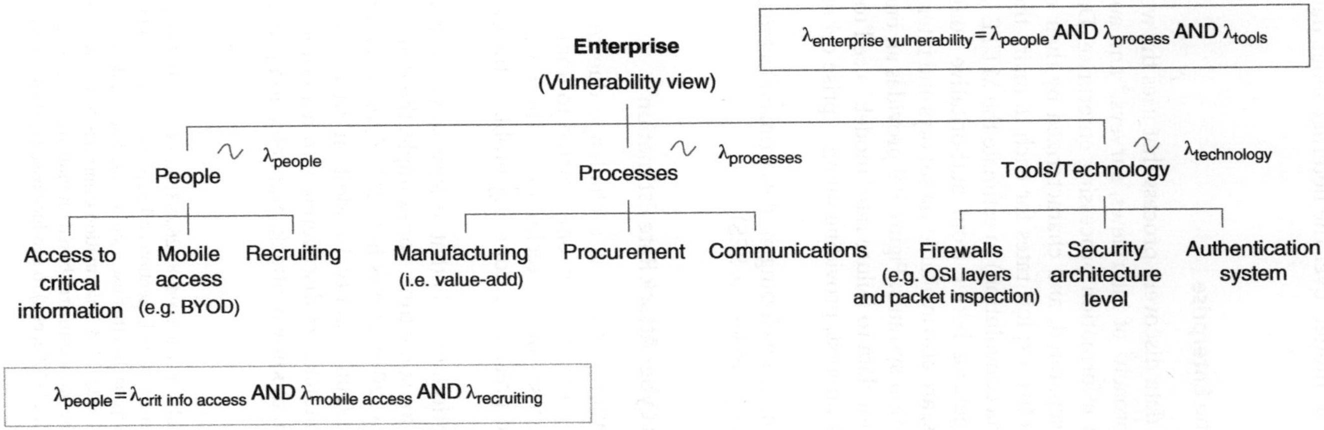

Figure 9.8 Enterprise model and parameterization.

in an "additive" combination of failure rates, leveraging the Palm-Khintchine Theorem, to develop a "model" over the heterogeneous data for the respective domains.

9.3.2 Understand the Enterprise

Figure 9.7 provides a data discovery process that uses the well-known knowledge-engineering approach[2] of interviews, surveys,[3] and automated tools to collect data about an information processing enterprise. Data is then structured into domains, separated, and characterized by their estimated exploit rate. Given the respective exploit rates for each domain, these "models," are then combined (e.g. via convolution) to estimate the MTTE (t_{MTTE}).

Cyberattack data, getting better from authoritative sources in terms of attacks/vulnerabilities, are also available via surveys and interviews for tailored evaluation of information systems. Figure 9.9 provides an end-to-end process for translating interview data to failure rate "models" used to estimate vulnerability for a system of interest, providing an enterprise cyber risk estimate at one instant in time.

Figure 9.9's approach, in developing models from collected data, is organized via an entity-relationship model (i.e. SES).

9.3.3 Sampling and Cyber Attack Rate Estimation

Estimating a distribution for t_{mtte} is a challenge due to the lack of data. Manufacturing quality, for example, only started to improve after the adoption of statistical process control techniques. The opacity of cyberattack/exploit data therefore presents a counting problem for developing model-based approaches.

An ongoing assumption, here, is that at least some cyberattacks are undetected, and can go on, for some time. For example, the Ponemon Institute (The Ponemon Institute, LLC 2014) shows how 2/3 of system administrators have a challenge defining risk/vulnerability. In addition, Mandiant (2014) estimates up to 250 days before attacks are discovered within a system. We are therefore dealing with a random, unknown, attack, and detection process.

2 ABB's "Cyber Security Analyzer" patented (8,726,393) on 13 May 2014 and McAfee's "System and Method for Network Vulnerability Detection and Reporting" patented (8,700,767) on 15 April 2014 are recently patented approaches that use a similar method. Bank of America also has a patent application (20130067581) titled "Information Control Self Assessment" along these lines.

3 One approach is to provide questionnaires based on industry standard policies (e.g. SANS 20, ISA-99, NERC-CIP, ISO 27001/27002, etc.) to produce baseline data concerning an enterprise's current security posture.

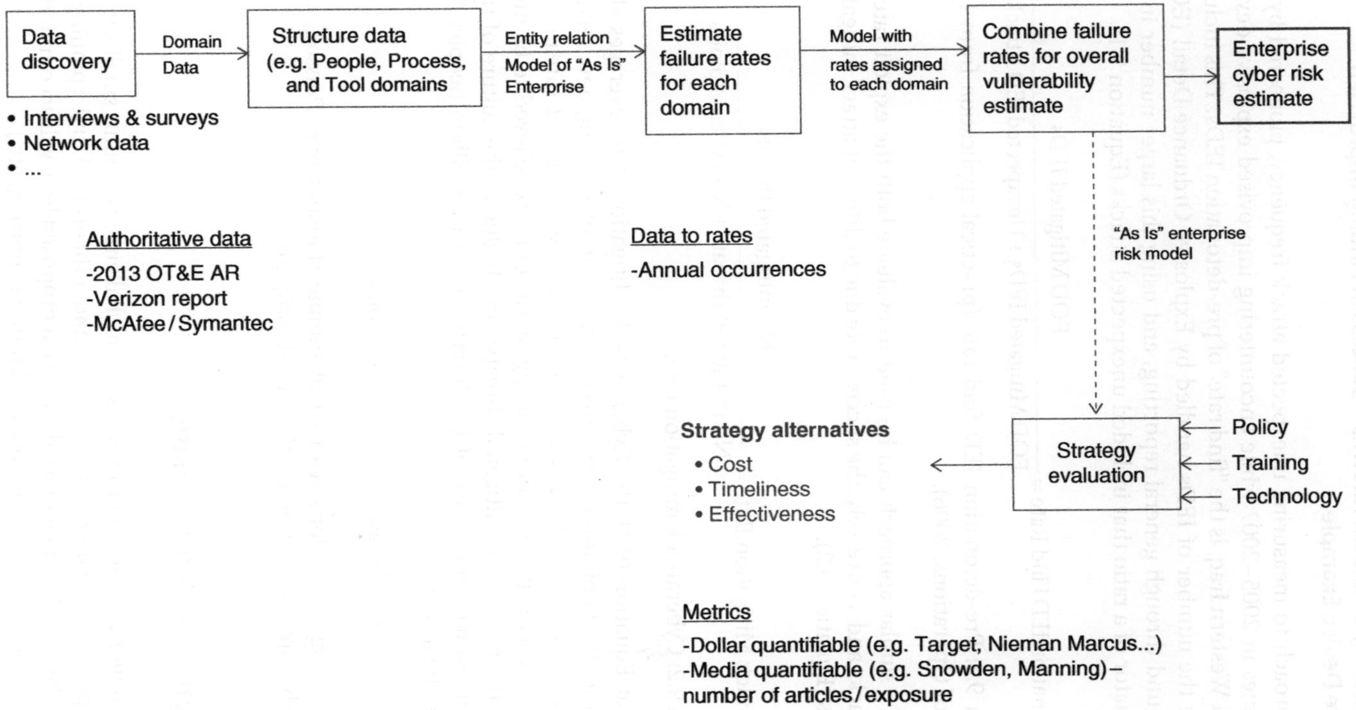

Figure 9.9 Data to strategy evaluation.

9.3.4 Finding Unknown Knowns – Success in Finding Improvised Explosive Device Example

One approach to measuring unexpected attack frequency, pioneered by the US Marines in 2005–2007 while encountering improvised explosive devices (IEDs) in Western Iraq, is the "find rate" of pre-detonation IEDs. This includes counting the number of IEDs handled by Explosive Ordnance Detail (EOD) teams found through general reporting, and using this larger number in the denominator of a ratio that included unexpected attacks (Equation 9.1).

$$\text{Pre-Detonation IED Find Rate} = \frac{\text{EOD Mitigated IEDs}}{\text{EOD Mitigated IEDs} + \text{Unexpected IED Attacks}}$$

Equation 9.1: Pre-detonation IED find rate (practical application from Iraqi Theater of Operations, 2006).

I believe a similar approach can be used to evaluate both the exploit rate for cyberattacks and, conversely, the success rate due to different attack mitigation strategies (Equation 9.2).

$$\text{Cyber Attack Mitigation Rate} = \frac{\text{No. mitigated threats}}{\text{No. mitigated threats} + \text{No. cyber exploits}}$$

Equation 9.2: Cyberattack mitigation rate.

Looking at Equation (9.2), the Cyber Attack Mitigation Rate leverages all of the general effects of annual employee training, implementing new security processes, software patching, etc. in the number of mitigated threats. The number of cyber exploits, more challenging to count, is believed to be much less than the number of mitigated threats; simply due to the culture of trust that results in day-to-day use of the Internet by over 1 billion people. This leads us to believe that

$$\text{No. mitigated threats} \gg \text{No. cyber exploits}$$

Describing cyber exploits with small sample statistics (e.g. Poisson, etc.) therefore becomes an option for this initial estimate.

9.4 Cyber Exploit Estimation

While sample sizes for cyber exploits are a challenge to count, small sample statistics provide us with baseline models, and a theoretical underpinning, to start our inquiry. The model-based approach proposed here will accommodate any phenomenological description (e.g. binomial, normal, etc.).

Data on cyberattacks is a challenge to get. Commercial attacks (e.g. Target, Nieman Marcus, etc.) make the news while the average consumer buys books on-line, uses her credit card multiple times per day, and logs on to the Internet for both business and personal use.

In addition, most of the respected data sources on cyberattacks consist of reports or surveys. An example of the current state of available data is a recent report on Insider Threat (The Ponemon Institute, LLC 2014). The report focuses on privileged users (e.g. DB Admins, Network Engineers, etc.) using their assigned permissions beyond the scope of their assigned role. Data from 693 respondents, collected on a Likert scale, resulted in percentages – an example being "47% see social engineering of privileged users as a serious threat for their network's exploitation ..." While this is good, general, information, it lacks the granularity required for developing behavioral models that can be used for prediction and model-based control; models with established base rates from available samples.

Seeing the human as a key component of any situational awareness evaluation, one approach to SA is shown in Table 9.6.

Combining Table 9.6's insight with a model-based approach provides a wrap-up technique for evaluating overall risk, in potentially real-time, to match the Δt of the attack surface. This includes using the rare event nature of exploits and convert this into an exploit rate for people, processes, and technology. Leveraging the Poisson distribution, for this initial estimate, the probability of exploit over the time period $(t, t + \mathrm{d}t]$, is $\lambda \mathrm{d}t$, and the probability of no exploit occurring over $(t, t + \mathrm{d}t]$ is $1^\circ - {}^\circ\lambda \mathrm{d}t$.

Next, consider our outcome variable *Yt* as the number of events, or exploits, that have occurred in the time interval of length h. For such a process, the probability that the number of exploits occurring in $(t, t + h]$ is equal to some value $y \in \{0, 1, 2, 3, \ldots\}$ is:

$$\Pr(Y_t = y) = \frac{\lambda h^y \left(e^{-\lambda h}\right)}{y!}$$

Equation 9.3: Poisson process.

Table 9.6 Situational awareness (SA) – levels and indicators.

Level	Time period	Indicator(s)
Strategic	Months – quarters	Anomaly detection
Tactical/Operational	Weeks – months	Process adherence
Technical	Hours – days	SIEM monitoring

Equation (9.3) is what is known as a Poisson process; events occur independently with a constant probability equal to λ times the length of the interval (that is, λh).

For a large number of Bernoulli trials, where the probability of an event in any one trial is small, the total number of events observed will follow a Poisson distribution. In addition, the "rate" can also be interpreted as the expected number of events during an observation period t. In fact, for a Poisson variate Y, $E(Y) = \lambda$. As λ increases, several interesting things happen

1) The mean/mode of the distribution gets bigger.
2) The variance of the distribution gets larger as well. This also makes sense: since the variable is bounded from below, its variability will necessarily get larger with its mean. In fact, in the Poisson, the mean equals the variance (that is, $E(Y) = \mathrm{Var}(Y) = \lambda$).
3) The distribution becomes more Normal-looking (and, in fact, becomes more Normal).

While the Poisson process provides a flexible baseline to initially model our exploit rates, we use Figure 9.10, expanding on Figure 9.4, to provide an entity-relational structuring for each domain's descriptive data (e.g. exploit rates).

9.4.1 Enterprise Failure Estimation due to Cyber Effects

Figure 9.10's SES, similar to a SysML structure, is used for enterprise design (Couretas 1998a, b) and is a general structuring that provides both visual enterprise decomposition and a means to organize system data described by any distribution.

Exponential Distribution provides a rough approximation to Enterprise security failure. Advantages include:

- Get the conversation started about enterprise security structure (i.e. SES of Enterprise).
- Initial cut at Enterprise risk model (i.e. more accurate approaches available as data quality increases).
- Leverage well-known mean time to failure for cyber domain.
- Cyber Risk Estimation = Mean Time to Exploit (MTTE)

$$\mathrm{MTTE} \sim 1 - e^{-\lambda_{\text{vulnerability}} * t}$$

Equation 9.4: Mean time to exploit (MTTE.)

Figure 9.11's example uses 50% as an example marker; providing modeled enterprise with an approximately 2-month MTTE.

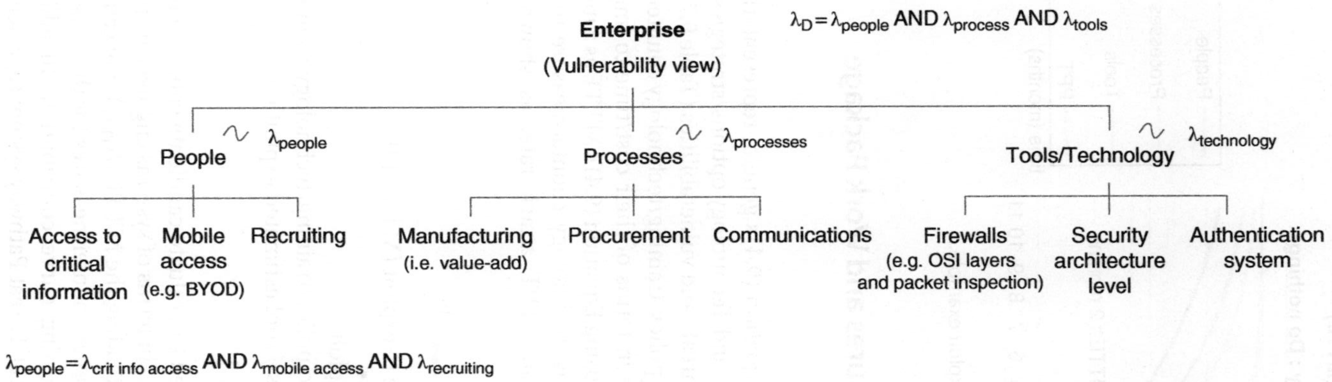

Figure 9.10 Enterprise vulnerability – exploit rates by domain.

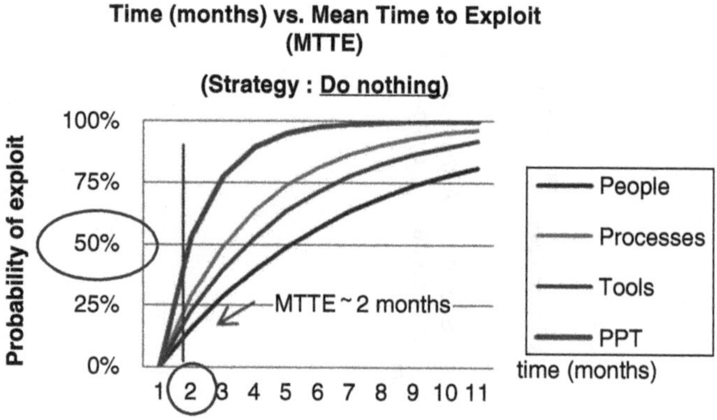

Figure 9.11 Mean time to exploit example.

9.5 Countermeasures and Work Package Construction

One of the benefits of Equation (9.4)'s general representation is the ability to evaluate Policy, Training, and Technology options/strategies as potential countermeasures to system threats and vulnerabilities (Table 9.7).

As shown in Table 9.7, policy/training/technology countermeasures are relatively simple to evaluate in terms of their cost/time to implement, and their estimated effectiveness, using Equation (9.4)'s MTTE as a measure of effectiveness. The Policy/Training/Technology countermeasure therefore becomes a discount on the estimated PPT attack rate, as shown in the following equation:

$$\text{MTTE} \sim 1 - \text{e}^{-\left(D^{*}\lambda_{\text{vulnerability}}\right)^{*}t}$$

Equation 9.5: Mean time to exploit (MTTE)(2).

MTTE: Mean time to exploit
D: Discount factor due to policy/training/technology countermeasure
$\lambda_{\text{vulnerability}}$: People/Process/Tool estimated exploit rate
t: Time

This model is then used to evaluate candidate countermeasures (e.g. policy, training, technology, etc.) in terms of systems engineering measures (e.g. time, quality, cost, etc.) for extending the MTTE (t_{MTTE}). We can use this construct to evaluate work packages as countermeasures for their discrete (i.e. one-time fix) and continuous learning curve contribution. In addition, this can be complemented by a look at Threat Pathway system representation (Nunes-Vaz

Table 9.7 Example countermeasures as work packages.

Packages/domain and work package		Cyber enterprise domain affected by work packages			Work package time/ cost estimate	
Work packages		People (λ_{people})	Process ($\lambda_{process}$)	Tool (λ_{tool})	Implementation time	Cost ($ K)
Policy	Access	•	∘	∘	Months	10's
	Mobile device	•	•	•	Months	10's
	Critical information	•	•	∘	Months	10's
	Phishing	•	∘	∘	Weeks	10's
Training	Internet use	•	∘	∘	Weeks	10's
	Social engineering	•	•	∘	Weeks	10's
	Firewalls	∘	•	•	Days	100's
Technology	M&C	∘	∘	•	Days	100's
	Authentication	•	∘	•	Weeks	100's

et al. 2014), and subsequent threat maturity estimation (e.g. using the beta distribution to estimate either threat or countermeasure development time), as an approach for estimating emerging threats.

9.6 Conclusions and Future Work

Using the SES-based system representation, we can evaluate current people/process/tool remedies for given threats and determine how to segment an enterprise description using layered representations (e.g. deter, deny, etc.) with MTTE for pre-event risk. In addition, we can extend this approach to post-event considerations that include evaluating resilience and subsequent mean time to repair, along with an overall mission system Availability estimate with both pre- and post-event risk evaluation.

Future work, therefore, includes both technical and management tools for evaluating systems for cyber security. Model-based approaches capture management best practices in recently developed technologies. One example includes the expanding on current security information and event management (SIEM) systems with automatic controls' responses, similar to what Canada's DRDC is developing with their ARMOUR framework (DRDC (Canada) 2014a, b). Management evaluation of cyber systems includes expanding on MTTE-based approaches for getting an architectural view of the current landscape of cyber risks and estimating how to systematically optimize the cost of security investments.[4]

Real-time approaches, such as streaming analysis (Streilein et al. 2011), develop and the discrete distributions that describe patterns of behavior, using them as a form of a model to perform anomaly detection for the enterprise. Table 9.8 provides two examples of deep packet inspection platforms.

Each of the architectures in Table 9.8 include three layers in developing greater resolution for their inspection activities. These approaches would be complemented by using a model-based approach that decomposes the enterprise into domains, with each domain represented by probability distributions, lending the overall system to the implementation of model-based control. State observers (Luenberger 1979), often used in the energy domain, are implemented via "Luenberger Observers" to monitor natural gas line networks for detecting and reporting on anomalous pressure changes, assisting operators in pinpointing the location of a leak or breakage in a gas line network that can easily span thousands of miles. A similar approach can be used for cyber.

4 This could take the form of a tool, developed in software, that includes both failure rate distributions and the effects of policy, training, and technology countermeasures.

Table 9.8 Deep packet inspection platform examples (Einstein and SORM).

System	Objective	Country	Source
System for Operative Investigative Activities (SORM)	Monitor all communications around the Sochi Olympics (e.g. including deep packet inspection for monitoring/filtering data) **SORM-1** intercepts telephone traffic, including mobile networks; **SORM-2** monitors Internet communication, including VoIP (Voice over Internet Protocol) programs like Skype; and **SORM-3** gathers information from all types of communication media.	Russia	http://en.wikipedia.org/wiki/SORM http://www.theguardian.com/world/2013/oct/06/sochi-olympic-venues-kremlin-surveillance http://themoscownews.com/russia/20130671/191621273.html
EINSTEIN	EINSTEIN is an intrusion detection system (IDS) for monitoring and analyzing Internet traffic as it moves in and out of United States federal government networks. EINSTEIN filters packets at the gateway and reports anomalies to the United States Computer Emergency Readiness Team (US-CERT) at the Department of Homeland Security. **Einstein 1** (2006) – analyzes network flow information to provide a high-level view for observing potential malicious activity. **Einstein 2** (2008) – automated system that incorporates intrusion detection based on custom signatures of known or suspected threats. **Einstein 3** (2013) – detects malicious traffic on government networks and stops that traffic before it does harm.	USA	http://gcn.com/Articles/2013/07/24/Einstein-3-automated-malware-blocking.aspx?Page=2 http://searchsecurity.techtarget.com/definition/Einstein

While employing a state observer at the network level is currently an aspirational goal, the work here provides a baseline in terms of

1) representative distributions for exploit phenomena
2) a method of organizing known problem data via entity-relational hierarchy and
3) a technique for combining the respective domain descriptions into a single, strategic, measure for whole of enterprise evaluation.

Leveraging contemporary systems and software tools, including MBSE, DEVS, and classical probability to provide a state observer for monitoring and controlling multiple attack surfaces will be a step forward in overall cyber security. For example, manufacturing quality control, once a problem in the "too hard" bucket due to the multiple people/process/technology dimensions in the physical domain, is now a solved problem. Similarly, this approach provides a path forward for managing and controlling the multiple people/process/technology dimensions in the logical domain.

9.7 Questions

1 How do the standard four risk approaches (e.g. avoid, accept, transfer, and mitigate) apply to cyber?

2 What additional domains (i.e. beyond People, Processes and Technology (PPT)) help with demonstrating a "complete" attack surface?

3 How are mobile and dormant devices included in an attack surface description?

4 Why does Equation (9.4) work for current cyber terrain?

5 What does the FAIR Institute's Factor Analysis of Information Risk Model cover, in terms of the protected system?

6 Why are small sample statistics adequate for cyber modeling? What are the drawbacks?

7 How might work packages be constructed for countermeasure estimation in defensive cyber planning?

10

Cyber Modeling & Simulation (M&S) for Test and Evaluation (T&E)

Multiple cyber ranges and test beds are currently being used for testing and training (Davis and Magrath 2013), with more coming on-line each month, often for educational purposes. Each of these ranges includes a modeling platform for representation of IT systems of interest. OPNET and Exata are key names in producing cyber M&S development platforms. For example, the Joint Communication Simulation System (JCSS), which leverages OPNET, is commonly used for long-haul communication evaluation. Similarly, CORONA (Norman and Christopher 2013) uses OPNET for T&E experiments.

10.1 Background

The speed and combinatorial nature of the evolving cyber threat demands a more flexible modeling & simulation (M&S) approach using cyber ranges for system test and evaluation (T&E). One approach is to review the baseline process to conduct cyber-range events. This includes the logical range construct, that provides event environments to be constructed in a location-independent manner, along with its application to the various phases of cyber range events. We also describe how cyber M&S is used in cyber-range events, and the need for using more intelligent and autonomous simulations in the event control plane.

The Department of Defense (DoD) (Defense Science Board 2013) is keen to ensure that fielded military systems are hardened to cyber exploits and resilient to cyberattacks. Hardening systems requires T&E of the processes and technologies related to countering cyberattacks. Training personnel is an essential component of making systems resilient to cyberattacks. The cost of using actual cyber assets or systems for T&E and training can be prohibitive in large scale, and therefore cyber M&S is used to reduce cost while doing cyber-range events for testing and training. The foundational concepts for using cyber M&S in cyber-range events are introduced here, along with applicable definitions and constructs.

An Introduction to Cyber Modeling and Simulation, First Edition. Jerry M. Couretas.

The Chairman, Joint Chiefs of Staff, issued an instruction in February 2011 (Joint Chiefs of Staff 2014) mandating that all DoD exercises begin to include realistic cyberattacks into their war games. To paraphrase the threat,

> If this level of damage can be done by a few smart people, in a few days, using tools available to everyone, imagine what a determined, sophisticated adversary with large amounts of people, time, and money could do.

Multiple efforts were started to coordinate cyber training, T&E. For example, the Joint Mission Environment Test Capability (JMETC) (Arnwine 2015) has the following goals for Cyber T&E:

- T&E must accurately and affordability measure cyberspace effectiveness and vulnerabilities of DoD systems (DoD Operatonal Test and Evaluation [DOT&E] 2015).
- Address the requirements for building Cyberspace T&E Process, Methodology, Infrastructure, and Workforce.

Cyber-range events vary in complexity and their objectives, and cover a broad spectrum of event types. For example, some events are conducted for training cyber protection forces, and some are conducted for evaluation of people, process, and technology through large-scale exercises, and yet others are conducted for developmental testing (DT) or operational testing (OT). Events may also be conducted for experimentation with technology or tactics, or to assess mission readiness. An early example of a cyber-range used for academic studies is the DETER test bed (The DETER Testbed: Overview 2004). DoD Enterprise Cyber Range Environment (DECRE (DOT&E 2013)) is a federated set of ranges within DoD, comprised of the National Cyber Range (NCR), the DoD Cyber Security Range, the Joint Information Operations Range (JIOR), and the Joint Staff J6's C4 Assessments Division (C4AD) (Table 10.1).

Table 10.1's cyber ranges are used to support cyber-range events such as Cyber Flag (Hansen 2008). In addition, the need for interoperability is being facilitated by the Cyber Range Interoperability Standards Working Group (CRIS WG) (Damodaran and Smith 2015).

10.2 Cyber Range Interoperability Standards (CRIS)

While Table 10.1 provides an overview of current cyber ranges and their capabilities within the DoD, standards for the interoperability of the cyber-range events are still developing. Currently, the ground work is still being laid in terms of coordination bodies and authoritative data common to professional fields. The primary use cases considered by the standards effort are

Table 10.1 Current US government cyber ranges.

Cyber range	Management	Function
National Cyber Range (NCR)	Joint Mission Environment Test Capability (JMETC)	General cyber system evaluation capability for large-scale scenario testing (Pathmanathan 2013) – transitioned from DARPA in October, 2012.
Cyber Security Range (CSR) (MANTECH 2018)	Defense Information Systems Agency (DISA)	Evaluation of bandwidth effects on operations
C4AD	Joint Staff J6 Test Resource Management Center (TRMC) (Ferguson, 2014)	Interoperability assessments, technology integration, and persistent C4 environment
Joint IO Range (JIOR)	Joint Staff J7	Nation-wide network of 68 nodes for live, virtual, and constructive operations across the full spectrum of security classifications (National Defense Authorization Act (NDAA) for Fiscal Year 2013, 2012)

cyber-range events, including training, exercises, and testing events. The reasons behind considering these use cases initially is that first, they represent a majority of the events, and second, the other use cases, such as mission rehearsals, are expected to be supported by the same standards with minor changes. A prominent standards effort is the CRIS WG (Damodaran and Smith 2015) that was formed by TRMC in collaboration with MIT Lincoln Laboratory in 2012 to foster interoperability of tools used to support cyber-range events such as Cyber Flag (Hansen 2008), which developed a cyber-range lexicon (Damodaran and Smith 2015), and a baseline process for cyber-range events (Bakke and Suresh 2015).

10.3 Cyber Range Event Process and Logical Range

The purpose of cyber-ranges is to conduct events. The events that are conducted will follow a baseline process, and therefore, it is beneficial to explore this process. While the overall flow of a cyber-range event (Figure 10.1) follows a familiar sequence, the incorporation of a logical range provides for both event scalability and the opportunity to add scope, via the incorporation of other systems, to the T&E or training event.

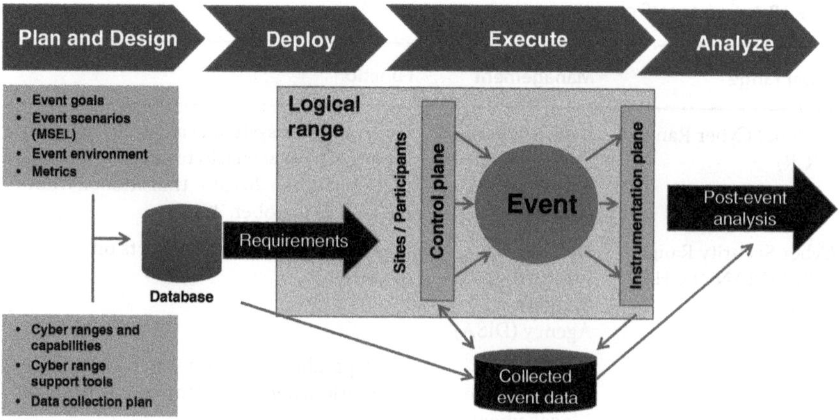

Figure 10.1 Cyber event process overview.

Figure 10.1 describes the overall flow of a cyber-event. At the top of Figure 10.1, the four distinct phases for most cyber-range events: plan, deploy, execute, and analyze are shown. Activities in the planning phase, shown below "Plan" (Figure 10.1), include gathering and specification of event requirements such as event goals, design of all aspects of the event, and scheduling the event. The outputs from the planning phase are captured and stored for use in the remaining phases. A full-fledged planning phase may not exist for some events that have a "persistent event environment" that is planned once, and subsequently deployed as many times as necessary using the event configuration of a previous event. It is normal for an event to iterate through one or more phases multiple times before proceeding to the next phase.

The deployment phase includes the deployment of the logical range (Figure 10.2). The logical range construct is a useful abstraction, and it has an implementation. The logical range provides for both event scalability and the opportunity to add scope, via the incorporation of other systems, to the T&E or training event. The cyber capabilities or infrastructure that are required for an event may not be available at a single site of a cyber-range, and therefore, multiple sites of one or more cyber-ranges may need to be interconnected to support an event. Alternatively, a single cyber-range may be supporting multiple events simultaneously. The logical range construct is used to abstract away the details of where the range infrastructure elements, including specific assets and capabilities, are physically located. The logical range is created based on the requirements and designs developed during the planning phase. The logical range is self-contained and is isolated from its surroundings. This isolation serves two purposes. First, the technologies that are deployed within the logical range do not escape into the real-world causing damage. Second, this isolation prevents unintended exposure of the logical range.

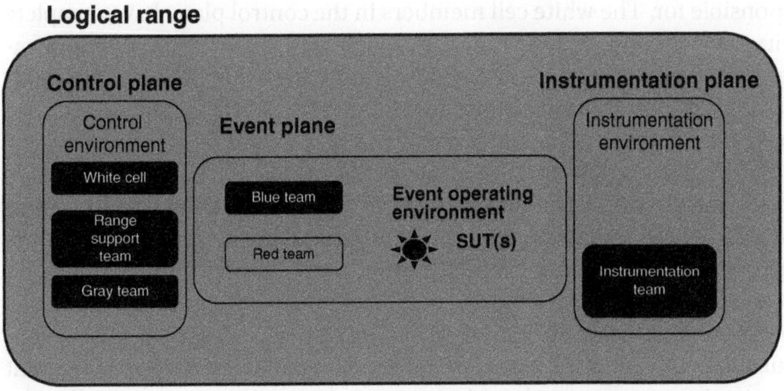

Figure 10.2 Logical range and the control, event, and instrumentation planes.

A cyber-range event is conducted using representative and operationally relevant cyberspace elements. Cyberspace is defined as "A global domain within the information environment consisting of the interdependent network of information systems infrastructures including the Internet, telecommunications networks, computer systems, and embedded processors and controllers" (CNSSI 2010). Therefore, cyberspace spans every system that has cyber elements, irrespective of whether a system is connected to a network. For example, a network router is in cyberspace, and so are all avionics components of an airplane. An operationally relevant and representative version of the actual environment with any additional features, including potentially one or more System-Under-Test (SUT), are deployed (Figure 10.2) on the logical range. This environment is called the event operating environment. The event plane includes the blue and red participants of the event who operate within the event operating environment as well as the operating environment. Since all participants in an event occupy the logical range, the logical range extends to, and includes, the remote terminals of any remote participants.

A control plane is deployed in the logical range for controlling the execution in the event operating environment. The participants in an event are organized in teams or cells. The control plane includes participants from all the teams operating on the control environment. A range support team, sometimes also called a green team, is responsible for the correct operation and availability of the infrastructure used to build the logical range. This team monitors the infrastructure, ensures its healthy operation, and makes any changes requested by the white cell on the infrastructure. The gray team is responsible for generating traffic within the operating environment. Simulated traffic provides the necessary background traffic and activities to conduct an event successfully. This team also monitors the health and status of the traffic generation systems they are responsible for. The white cell members in the control plane have complete authority on the entire logical range, participants, and the underlying infrastructure to make any changes at any time during the event. The instrumentation team installs appropriate probes in the operating environment, and is also responsible for the collection and archival of event data.

During the planning phase, several items are specified. These items include event goals, metrics, infrastructure required to support the event, scenarios to be executed during the event, the various environments of the event, cyber assets and capabilities needed for the event, the cyber ranges participating in the event, and tools required to support the event. The planning phase also results in the creation of plans and schedule for the various event milestones, asset and capability acquisition, as well as the plans for collecting data. The designs of the event operating environment, control and instrumentation environments, as well as the Mission Scenario Event List (MSEL) are also finalized during the planning phase. In the deployment phase, the elements of the event environment are deployed, verified, and validated. The execution of the event occurs during the execution phase, and event data is generated, and collected during the execution. The event data is used for monitoring the progress of the event as well as for analyzing the effectiveness of the event and report the event results to stakeholders.

The duration of event execution phase, typically a few days to a couple of weeks, is determined during the planning phase. The planning phase includes the specification, design, and scheduling of all activities related to the logical range, event operating environment, and the event phases. The planning phase may take anywhere from a few weeks to several months depending on how well the requirements are defined, ease with which the right assets and capabilities are acquired, reserved, and scheduled for the event, and the design complexity of the event environment. The activities during the planning phase are usually not conducted on the logical range. The bulk of these activities are conducted in a planning network prior to the construction of the logical range. However, sometimes updates to any of the plans or designs are done within the logical range in the control plane. The planning network can be the standard enterprise environment used for day-to-day business. The deployment phase may be a few days or months depending on the scale and complexity of the logical range, environment, and how widespread and integrated the use of automation tools are. The analysis phase may take a few days to months depending on the amount of event data generated by the event.

10.4 Live, Virtual, and Constructive (LVC) for Cyber

The event operating environment contains actual cyber assets and capabilities, or their models. The Live, Virtual, and Constructive (LVC) terminology (Henninger 2008a, b) used to classify kinetic-range simulations can be used in cyber ranges as well, though these terms need re-definition within the cyber context. In cyberspace, LVC simulations are used in events to provide the appropriate level of fidelity required for the event operating environment. Below, we define the LVC terminology briefly. For a more descriptive definition of these terms, please refer to (Damodaran and Smith 2015).

We assume the definitions of the terms Model and Simulation (Zeigler et al. 2000), and Emulation (SISO, 1999). A real-world asset or system used to represent that asset or system, respectively, in a simulation is labeled as an "actual" asset or system. When a simulated model of an asset or system provides some of the operationally relevant interactive interfaces, protocols, or features of the actual asset, including partial or complete simulation of internals of an asset or system, it is called a "representative" model. It is possible to conduct emulation over the provided interfaces or protocols. When actual assets interact with representative models of protocol-level fidelity representations of actual systems, where ease of (re)configuration, replication, restoration, and physical limitations make a representative system preferred over the actual one, there is no physical representation of the actual system. In this specific case, a representative model of the system only provides a cyber "attack surface," through the protocol/packet interfaces, but not asset internals or attributes susceptible to other attacks. An agent that embodies the behavior of a human is a representative asset model, as long as the agent is capable of interacting effectively with actual assets or systems, or their representative models.

Some of the assets or systems used in simulations may not provide any of the operationally relevant interactive interfaces, or protocols, of the actual asset or system. Such a system or asset model is referred to as "limited." A limited model may not provide relevant attack surfaces or participate in synthetic traffic generation. A model does not qualify as either limited, or representative model for an event if it does not provide any operationally relevant interfaces, protocols, or features for that event.

In looking at the differences among LVC cyber simulations using actual systems or assets or their representative, or limited models. Live (Cyber) Simulation: In this type of simulation, real-world assets operate on/with real-world systems and protocols. Even though actual assets are used, these are considered simulations because the scenario is simulated and attacks are not conducted against a live enemy. Examples include:

- Actual operators, actual network devices, actual machines, actual non-emulated/simulated software.
- Packet, protocol, or frequency-level attack and response launched by actual systems and/or live attackers.

Virtual (Cyber) Simulation: In this type of simulation, actual assets may interact with limited or representative system models, and limited or representative asset models may interact with actual systems.

Examples include:

- Asset emulators running on virtual machines.
- Automated response of a virtual machine to an attack.
- Replay of a logged live attack onto the live or virtual systems.
- Automated or semi-automated attack simulators that replicate the actions of a live red team or real-world threat.
- Simulated users (Assets) in a traffic generator using actual systems to generate traffic.
- Accurate (high-fidelity) representations of system administrator GUIs.

Constructive (Cyber) Simulation: In this type of simulation, limited or representative asset models interact with limited or representative system models. The simulated systems are characterized by lower fidelity global/enterprise-level networks and effects representations, and are not vulnerable to direct live or virtual exploits and manipulation.

Examples:

- Simulated internet-scale traffic generation, background noise, and high-volume gray-space.
- Virus infection and worm propagation simulations.
- Asset representations with simulation interfaces that must be translated or bridged to connect with virtual and live assets.

While the terminology of LVC simulation is useful for distinguishing among different types of simulations, it is possible that, in practice, a "federated simulation" is the result of interaction among multiple simulations, often of different types. In such a situation, the type of the federated simulation is the lowest of the simulation types of the interacting simulations, where "live" is the highest type and "constructive" is the lowest type. For example, let us say a federated simulation results from the interaction of virtual simulations and constructive simulations. The type of the federated simulation is then constructive.

10.4.1 Role of LVC in Capability Development

Let us explore the role of LVC simulations in capability development by analyzing the typical process of development and testing of a capability. Mathematical modeling or analysis of the capability and its operating environment is a usual first step. This analysis provides a highly scalable and comparatively inexpensive approach to quantitatively analyze the capability. However, mathematical analysis may abstract away many properties of the capability and the operating environment, and therefore utilizes limited models.

Constructive simulation permits the use of limited or representative models that have more features, and therefore, provides higher fidelity than the mathematical analysis, yet similar level of scalability and affordability. Using LVC simulation within a cyber-range event with actual, representative, or virtual models of assets and systems provides for better fidelity than a purely constructive simulation. Eventually, a prototype of the capability may be developed, and tested using either a cyber-range using actual models, or in an actual operational environment, depending on the type of capability.

The cyber-ranges are similarly used for developing and conducting experimentation, development, training, or other types of events using limited, representative, or actual models with LVC simulation. We explore this topic in more detail in the next section.

10.4.2 Use of LVC Simulations in Cyber Range Events

LVC simulations may be used for multiple purposes in a cyber-range event. Below, we discuss different uses for LVC simulations in a cyber-range event.

Most cyberspace events require traffic generation to simulate the dynamic environment that a system under test (SUT) would normally operate. There are multiple types of traffic generators with varying fidelity. The generated traffic may emanate or be consumed by any element, including the SUT, within the operating environment. A traffic generator supports both constructive and virtual simulations to generate the traffic. For example, in LARIAT (Rossey et al. 2002), a traffic generator developed by MIT Lincoln Laboratory, uses a model of the Internet, since actual Internet cannot be used for a cyber-range event. This Internet is used by agents with representative user models in a virtual simulation using actual applications and operating systems to generate background traffic. The fidelity of the traffic can be considerably improved by having the user agents organize themselves into communities and behaviors that mimic an actual operating environment, and with increased intelligent and autonomous traffic generation. In LARIAT, the Internet model does support application and network protocols such as HTTP, HTTPS, TCP/IP, SSH, and SMTP so that the websites and file servers in the Internet model can be accessed by the virtual user agents. The Internet may also be used by real blue or red participants in the event, and when that happens, the Internet traffic is generated as a virtual simulation, using representative and constructive models. One of the possible constructive models is that of the web site contents, while one of the representative models is that of the DNS service.

Traffic may also be generated through other means. For example, cyber elements removed from the original equipment and re-hosted in alternative hardware or software for convenient deployment within the event operating environment, and by extension, in logical range, may be fed previously recorded traffic from the actual system so that the cyber elements may interact with the rest of the operating environment.

Within the event operating environment, the interactions among the blue and red team members, and traffic generator(s) cause several changes in the operating environment. Some of these changes may be designated as "effects," or predefined anticipated changes to the event operating environment. These effects may be caused by traffic originated from another cyber element or participant in the event operating environment. An effect may also result from an explicit directive from the white cell. For example, a zero-day attack on a system may be substituted by the use of a surrogate causing the same effect as the zero-day attack. As another example, several systems may be turned off ("effect") to simulate the effect of a fire in the actual environment. Therefore, these effects may be simulated using appropriate models within the Control Plane and then the effect injected into the event plane. These models of effects may be simulated through virtual or constructive simulations.

It is often the case that while the blue and red teams may be interested in attacks on cyber key terrain, other attacks could be happening in the rest of the event operating environment. Simulation may be used to generate such attacks or attack missions on the event operating environment. For example, the Cyber Operational Architecture Training System (Morse et al. 2014a, b) is an example of providing range-based cyber effects into a command-level training exercise.[1]

10.5 Applying the Logical Range Construct to System Under Test (SUT) Interaction

The logical range construct described in the previous section can be applied fruitfully to analyze problems in events. We describe one such interesting problem, namely, how the elements in an operational environment should interact with a SUT during an event. Often, SUT will be deployed in the event operating environment. However, sometimes it may not be feasible to extend the logical range to include the SUT due to logistical or other reasons. In this case, an SUT is placed external to the logical range. For example, Table 10.2 provides an analysis of how an SUT may interact with an event operating environment based on both its deployment location and possible simulation approaches.

As shown in Table 10.2, when the SUT is in the logical range's operational environment, the SUT simulation state is embedded in the operational environment state; and the cause of an effect in the operational environment can be traced back to entities within the operational environment. In contrast, when the SUT is not in the operational environment of a specific logical range X (SUT may be in another logical range Y or not in any logical range), then, though the SUT simulation effects still correspond to the local operational environment state, entities within the operational environment lose their

1 As opposed to using red teams to attack the actual system of interest during an exercise.

Table 10.2 System Under Test (SUT) evaluation approaches.

Type of SUT	Location	Simulation approaches
Hardware	Logical range	1) Constructive simulation with limited model of the SUT in operational environment. 2) Virtual or live simulation with representative model or actual SUT in operational environment. 3) Live simulation with actual SUT in operational environment.
Software applications running on standard OS	Logical range	1) Constructive simulation with limited model of the SUT in operational environment. 2) Virtual simulation by re-hosting the application on a copy of the standard OS in the operational environment. 3) Live simulation with actual SUT in operational environment.
Software applications running on nonstandard OS	Logical range	1) Constructive simulation with limited model of the SUT in operational environment. 2) Live or virtual simulation with actual SUT in operational environment.
Hardware or Software application(s)	External to a logical range	1) Effects integration through control plane and instrumentation plane.

traceability to nonlocal effects. In this situation, the control plane has the responsibility to provide this information to the entities within the operational environment, as appropriate.

10.6 Conclusions

While M&S provides both a sanitary environment to test cyber effects and the requisite traceability for understanding threat behavior in an emulated attack, cyber M&S is still growing into the mature capability that current DoD stakeholders rely on in their current LVC simulations. In addition, the "right level" of cyber M&S is still being determined, in balancing the mix of actual assets and systems with limited and representative asset and system models for simulations.

Foundational concepts provided in this paper (e.g. logical ranges) (Figure 10.2) provide the tools, as shown in Table 10.2, for cyber M&S to grow beyond stand-alone simulations; coupling ranges in terms of both model composition and the communication of effects. Due to the inherently dynamic cyber terrain, consisting of both moving targets (Okhravi et al. 2013a, b; Colbaugh and Glass 2012), and necessarily dynamic defense, the foundational work described here provides an extensible M&S approach for a rapidly developing system of systems domain.

10.7 Questions

1 What are the differences between the SUT, event operating environment, and the control cell?

2 Why is LVC important for T&E?

3 Why would multiple individual cyber ranges be included in a logical range?

11

Developing Model-Based Cyber Modeling and Simulation Frameworks

While logical ranges provide the means for emulating systems of interest, M&S frameworks are still a necessary ingredient to provide the scale and scope to model contemporary cyber systems. Prescriptive approaches (Leversage and Byres 2007), for example, represent a system's specific behavior, and use process models, embodied in more detailed component-level descriptions (e.g. SysML), for more granular cyber system analysis. These structured approaches provide the "bottom up" behavioral rigor that composes cyber M&S frameworks.

11.1 Background

As early as 2001, the Modeling and Simulation Information Analysis Center (MSIAC) wrote a State of the Art Report (SOAR) (Waag et al. 2001) on Information Assurance (IA) to describe "cyber." Since then, we have seen IA transition from DIACAP to the current Risk Management Framework (RMF); each risk framework spanning the "bow-tie" with varying terms, depending on the domain that the risk evaluation approach was developed for. RMF, the current standard, is a flexible approach, that can be complemented by cyber M&S' dynamics to provide system-level analysis, such as attack surface estimation and operating metrics (e.g. Availability, Confidentiality, and Integrity).

11.2 Model-Based Systems Engineering (MBSE) and System of Systems Description (Data Centric)

MBSE is evolving from the software development community, currently being used for overall systems development. The Object Management Group's Systems Modeling Language (SysML) (Object Management Group [OMG] 2012), leveraging UML concepts, is an example of this extension from software to systems engineering (Figure 11.1).

An Introduction to Cyber Modeling and Simulation, First Edition. Jerry M. Couretas.

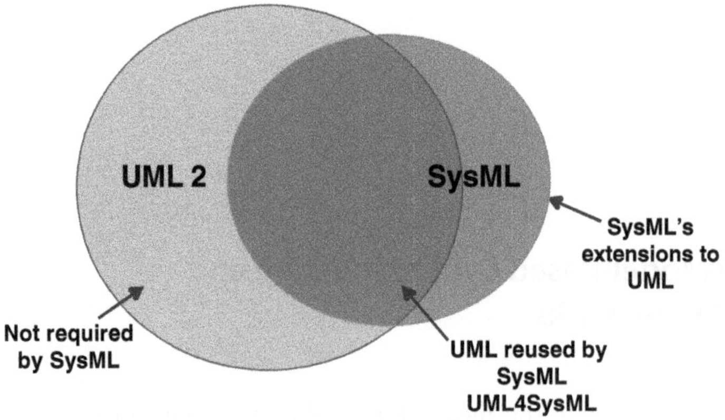

Figure 11.1 UML 2 and SysML.

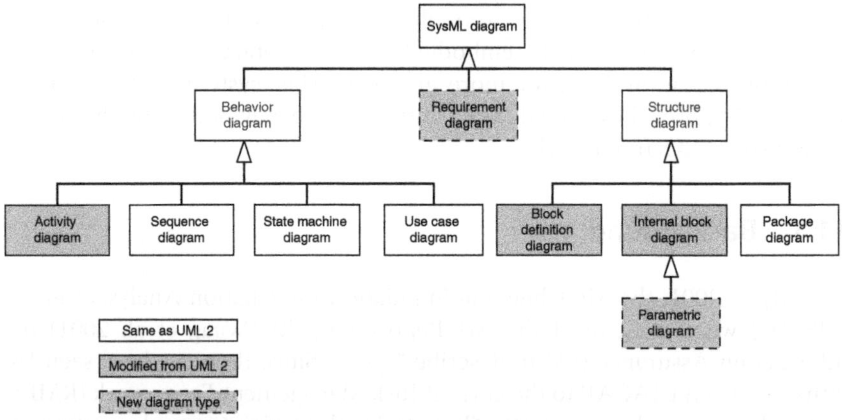

Figure 11.2 SysML diagram types.

SysML's top-down approach provides the Chief Engineer with a process to determine essential elements of data to support an assessment or decision (Estefan 2008). In addition, SysML's extensions of UML result in a broadly applicable systems level tool; for example, SysML diagram types shown in Figure 11.2.

In addition, SysML is intended to be compatible with the evolving ISO 10303-233 (Valilai and Houshmand 2009) systems engineering data interchange standard. Figure 11.2 provides an overview of existing systems engineering process standards and capability models.

11.3 Knowledge-Based Systems Engineering (KBSE) for Cyber Simulation

Knowledge-Based Systems Engineering (KBSE) provides a data-based approach for developing sophisticated training and simulation systems (Figure 11.3).

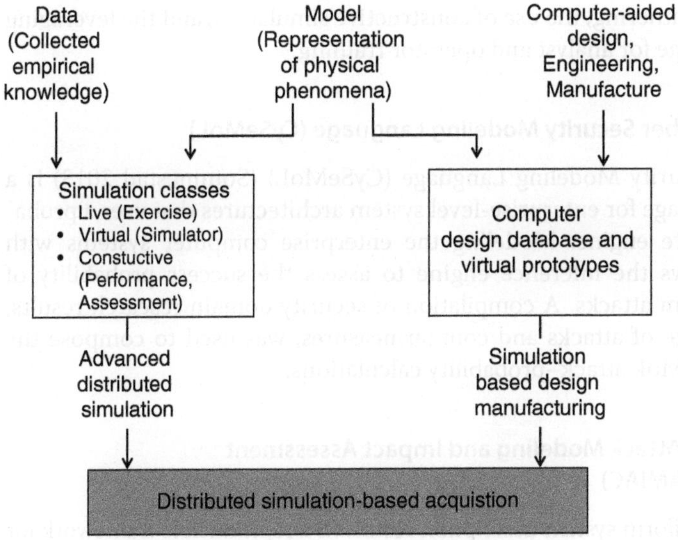

Figure 11.3 Legacy simulation-based acquisition development approach.

As shown in Figure 11.3, original concepts for simulation-based acquisition included leveraging computer-aided design (CAD) information, data at a level of modern KBSE. Combining KBSE with the behavioral simulation common to agent-based modeling has the potential for a comprehensive approach to cyber M&S.

11.3.1 DHS and SysML Modeling for Buildings (CEPHEID VARIABLE)

While DHS is known for its ICS-CERT's Cybersecurity Evaluation Tool (CSET®) for onsite assessments, network design architecture reviews, and network traffic analysis and verification, an additional development that we are seeing is the evolution of overall frameworks for cyber strategy formulation, leveraging threat behavior models for enterprise risk evaluation. An example of this more detailed modeling approach was a Small Business Innovative Research (SBIR) funded Cyber-to-Physical Domain Mapping Toolkit for Vulnerability Analysis and Critical Resource Identification Enablement (CEPHEID VARIABLE). Built by Knowledge Based Systems, Inc. (KBSI) (Small Business Innovative Research [SBIR] 2012), the goal of CEPHEID VARIABLE is to enable IT managers to perform vulnerability assessment of cyber–physical systems in an application framework that enables the acquisition, representation, storage, mapping, vulnerability, and dependency analysis of information linking cyber and physical resources; supporting both static and dynamic vulnerability analysis. This includes a leveraging of available, and evolving, NIST threat representations (Chapter 2, Table 2.1). As discussed in Chapter 8, cyber M&S includes requirements gathering, the use of constructive simulation, and the leveraging of this knowledge for analyst and operator training.

11.3.2 The Cyber Security Modeling Language (CySeMoL)

The Cyber Security Modeling Language (CySeMoL) (Sommestad 2013) is a modeling language for enterprise-level system architectures that uses a probabilistic inference engine. Modeling the enterprise computer systems with CySeMoL allows the inference engine to assess the success probability of candidate system attacks. A compilation of security domain research results, covering a range of attacks and countermeasures, was used to compose the corpus of CySeMoL attack–probability calculations.

11.3.3 Cyber Attack Modeling and Impact Assessment Component (CAMIAC)

Leveraging a uniform system description results in a reproducible framework for cyberattack modeling and impact assessment. A common approach to attack modeling and impact assessment is based on representing malefactors' behavior, generating attack graphs, calculating security metrics, and providing risk analysis procedures. The architecture of the Cyber Attack Modeling and Impact Assessment Component (CAMIAC) is proposed in (Kotenko and Chechulin 2013).

As shown in Figure 11.4, the CAMIAC authors present the prototype of the component, the results of experiments carried out, and comparative analysis of the techniques used.

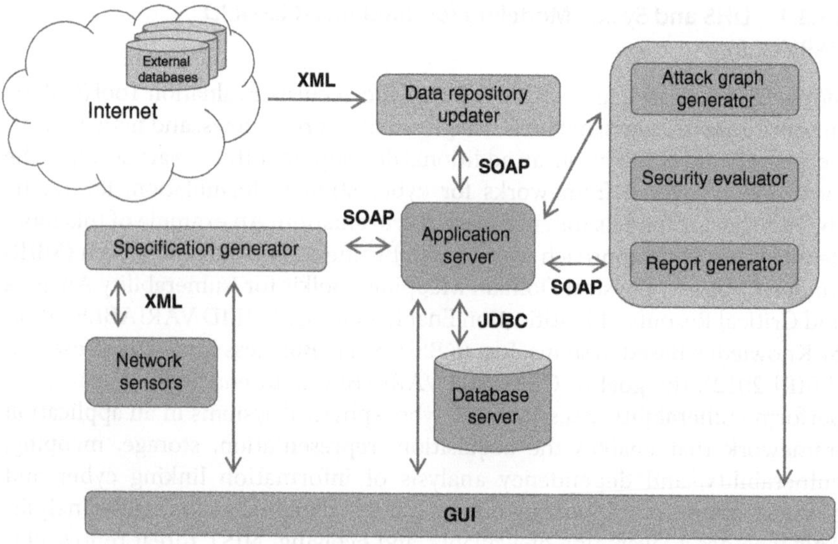

Figure 11.4 CAMIAC prototype architecture.

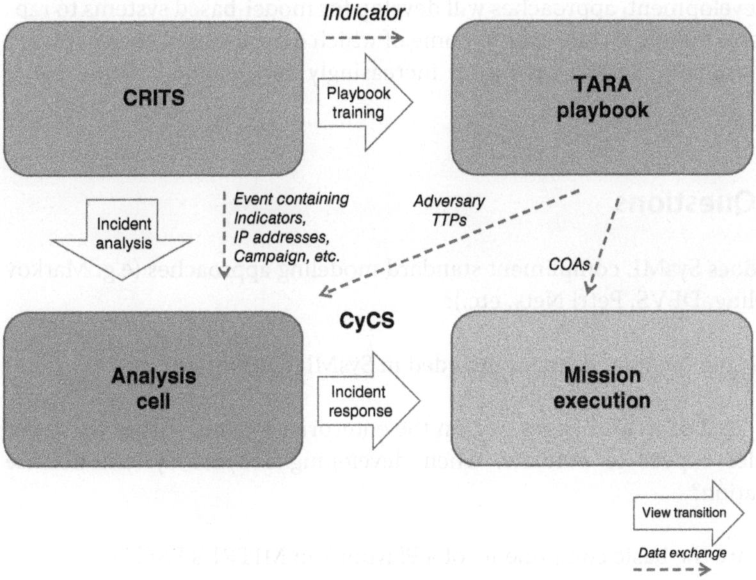

Figure 11.5 Federated Analysis of Cyber Threats (FACT) (MITRE 2015).

11.4 Architecture-Based Cyber System Optimization Framework

Modeling real cyber phenomena usually requires some level of abstraction (e.g. two player games, agents, etc.). MITRE's Collaborative Research Into Threats (CRITS) (Figure 11.5) provides an enterprise-level capability to "simulate" incidents over associated COAs and adversary TTPs.

Expanding on CRITS, Musman developed the Cyber Security Game (CSG) (Musman and Temin 2017), an approach and supporting software that quantitatively identifies cybersecurity risks and uses this metric to determine the optimal employment of security methods for any given investment level. CSG maximizes a system's ability to operate in today's contested cyber environment by minimizing its mission risk, where its risk measure is the inverse of its cyber resilience.

11.5 Conclusions

Contemporary modeling spans from data structuring (e.g. UML, MBSE, SySML…) to overall descriptive approaches (e.g. CySEMOL, CAMIAC), composing the foundation for a future of knowledge-based design in cyber systems. As these frameworks evolve to provide the foundation for secure cyber system development, approaches will develop for model-based systems to rapidly traverse potential state spaces, some of which may not have been explored by the designer, thereby providing increasingly secure and reliable cyber systems.

11.6 Questions

1 How does SysML compliment standard modeling approaches (e.g. Markov modeling, DEVS, Petri Nets, etc.)?

2 How is the "people" element included in SysML description?

3 What kind of state changes, within the enterprise system, should the cyber modeler expect to emulate, when developing a cyber system defense simulation?

4 What are the basic components of a Playbook in MITRE's FACT?

5 How does a Playbook overlap with a COA in MITRE's FACT?

12

Appendix: Cyber M&S Supporting Data, Tools, and Techniques

One of the challenges in developing engineering approaches for current cyber systems is the availability of principles-based building blocks, derived from empirical understanding, to be used as models. Table 12.1 gives examples of current accepted knowledge, well known in physical security, and potentially applicable to cyber, along with sources of both empirical understanding and developing ideas in the cyber domain.

Table 12.1's knowledge categories, initial taxonomic building blocks for cyber M&S, require additional system context, including cyber modeling considerations, for designing secure cyber systems.

12.1 Cyber Modeling Considerations

Cyber modeling leverages accepted knowledge from both physical security and computer-based system evaluations. For example, popular technology platforms, their availability, and required levels of expertise should be accounted for in the context of evaluating a cyber system. Evaluating technology via physical security analogs (e.g. Net Working Time (NWT) to access a safe) provides an opportunity for the modeler to validate candidate cyber models. This is especially useful if open-source cyber threat data complements model evaluation. While well-known critical security controls (CSCs) codify best practices for cybersecurity, developing evaluation approaches use situational awareness (SA) measures that are especially useful if the system/model is designed to automate a human task currently performed in a cyber environment (e.g. to detect or remediate threats).

12.1.1 Factors to Consider for Cyber Modeling

Factors that influence how a system is modeled are provided in Table 12.2 (Velez and Morana 2015).

An Introduction to Cyber Modeling and Simulation, First Edition. Jerry M. Couretas.

Table 12.1 Cyber M&S knowledge categories and examples.

Knowledge category	Example
Accepted knowledge	Net Working Time (NWT) (i.e. time to "crack" a safe)
	Psychological limits for Situational Awareness (SA)
	Critical Security Controls (CSCs)
Current empirical understanding	Threat data
	Market-proven tools currently providing value for cyber operators and professionals
Developing ideas	Published patents
	Patent applications
	Journal and conference publications

Table 12.2 spans the considerations for cyber threat modeling. While computer-based systems present attack surfaces that require combinatoric description, known practices in physical security use simple time estimates to abstract on access complexities (e.g. time to "crack" a safe).

12.1.2 Lessons Learned from Physical Security

Determining the burglary rating of a safe is a similar problem to determining the security rating of a network. Both involve a malicious threat agent attempting to compromise the system and take action resulting in loss. Safes in the United States are assigned a burglary and fire rating based on well-defined Underwriters Laboratory (UL) testing methodologies such as UL Standard 687. A few selected UL safe burglary ratings are given in Table 12.3.

The rating system is based around the concept of "Net Working Time" (NWT), the UL expression for the time that is spent attempting to break into the safe by testers using specified sets of tools such as diamond grinding tools and high-speed carbide tip drills. Thus, TL-15 means that the safe has been tested for NWT of 15 minutes using high-speed drills, saws, and other sophisticated penetrating equipment. The tool sets are also categorized into levels – TRTL-30 indicates that the safe has been tested for a NWT of 30 minutes, but with an extended range of tools such as torches. Assumptions about the processes include:

1) There is an implication that given the proper resources and enough time, any safe can eventually be broken into.

Table 12.2 Factors affecting time requirements for threat modeling.

Factor	Description	Example
Number of use cases	The number of actions that an application can perform as a result of a client request, scheduled job, or Application Programming Interface (API)	Actions that include buying items online, paying bills, exchanging content between entities, or managing accounts
Popularity of technology	The notoriety of a platform or software technology will provide attackers with the ability to have a sophisticated level of understanding on how to better exploit the software or platform	Any distributed servers, both open source and commercial
Availability of technology	The rarity of technology will affect probability levels of malicious users obtaining a copy of similar technologies to study its vulnerabilities for exploitation	Legacy software or proprietary software
Accessibility to technology	Cost of technology is not only a deterrent for legitimate, law-abiding companies, but also for those organizations that subsidize cybercrimes	Proprietary developed systems, kernels, or software
Level of expertise	Given that exploit scenarios move beyond the theoretical in application threat modeling, the appropriate level of expertise is needed to exploit vulnerabilities and take advantage of attack vectors. Depending on the level expertise, a threat modeler or team of security professionals may have varying levels of time constraints in trying to exploit a given vulnerability. This is very common and would require the security expert to be well versed in multiple talents to exploit vulnerable systems (Cho et al. 2016; Ben-Asher et al. 2015; Jones et al. 2015)	Experience with rare software/ platforms

Table 12.3 Selected Underwriters Laboratory (UL) safe burglary ratings.

UL rating	Net Working Time (NWT) (minutes)	Testing interpretation
TL-15	15	Tool Resistant (face only)
TL-30	30	Tool Resistant (face only)
TRTL-15X6	15	Torch & Tool Resistant (six sides)
TRTL-30X6	30	Torch & Tool Resistant (six sides)
TXTL-60	60	Torch & Tool Resistant

2) A safe is given a burglary rating based on its ability to withstand a focused attack by a team of knowledgeable safe crackers following a well-defined set of rules and procedures for testing.
3) The rules include using well-defined sets of common resources for safe cracking.
4) The resources available to the testers are organized into well-defined levels that represent increasing cost and complexity and decreasing availability to the average attacker.
5) Even though there might be other possibilities for attack, only a limited set of strategies will be used, based on the tester's detailed knowledge of the safe.

The UL rating does not attempt to promise that the safe is secure from all possible attack strategies – it is entirely possible that a design flaw might be uncovered that would allow an attacker to break into a given safe in seconds. However, from a statistical point of view, it is reasonable to assume that as a group, TL-30 safes are more secure than TL-15 safes. This ability to efficiently estimate a comparative security level for a given system is the core objective of looking Mean Time to Exploit (MTTE) (Chapter 9).

Learning from the safe rating methodology, MTTE for rating a network makes the following assumptions:

1) Given the proper resources and enough time, any network can be successfully attacked by an agent skilled in the art of electronic warfare.
2) A target network or device must be capable of surviving an attack for some minimally acceptable benchmark period (e.g. MTTE).
3) The average attacker will typically use a limited set of strategies based on their expertise and their knowledge of the target.
4) Attackers can be statistically grouped into levels, each with a common set of resources such as access to popular attack tools or a level of technical knowledge and skill.

Complementing the UL analog for cyber system access estimation is the cyber threat data providers' tactical and strategic outlook for current trends in system attack practices.

12.1.3 Cyber Threat Data Providers

Ponemon, Verizon, and Symantec are some of the most famous open-source cyber threat reports (Table 12.4).

Each of Table 12.4's threat data reports provides practical insights into current cyber operations. This includes informing the application of CSCs as a preventive measure for network security.

Table 12.4 Open-source cyber threat reports – organizations and missions.

Name	Mission
Ponemon	Ponemon Institute conducts independent research on privacy, data protection, and information security policy. Our goal is to enable organizations in both the private and public sectors to have a clearer understanding of the trends in practices, perceptions, and potential threats that will affect the collection, management, and safeguarding of personal and confidential information about individuals and organizations. Ponemon Institute research informs organizations on how to improve upon their data protection initiatives and enhance their brand and reputation as a trusted enterprise.
Verizon	The 2016 report continues our investigation into nine common threat patterns and how they are evolving from last year's report. The *2016 Data Breach Investigations Report* (DBIR) addresses several topics for the very first time: (i) What effect does mobile malware have on data security and (ii) How can you better estimate the financial impact of a data breach?
Symantec	The Internet Security Threat Report provides an overview and analysis of the year in global threat activity. The report is based on data from the Symantec Global Intelligence Network, which Symantec's analysts use to identify, analyze, and provide commentary on emerging trends in the dynamic threat landscape.

12.1.4 Critical Security Controls (CSCs)

CSCs (Table 12.5) are the product of multiple man-years of time and effort. Designed to be prioritized, based on their level of effective security, CSCs provide an easy reference for cybersecurity professionals, from beginning to advanced. An additional note about CSCs is that they are designed to be automated, being a first step in machine-based course of action (COA) reaction to cyber threats.

Some of Table 12.5's automated responses include patching, port closure, and packet screening (e.g. to find encrypted data transmission in an exfiltration). Similar to Table 12.5's controls are Australia's opposition force (Table 12.6), estimated to provide 85% of the network security requirements.

As shown in Tables 12.5 and 12.6, security controls provide a cross-institutional memory that can help security researchers approach key challenges to creating cyber agents that include (i) modeling the complex and continually evolving processes of cyber operations and (ii) leveraging the tools and data standards that enable cognitive agents to interoperate with networks invisibly to the user; distilling models of cyber offensive and defensive behavior based on observation and elaboration of human expertise.

12.1.5 Situational Awareness Measures

Cyber M&S benefits from a long line of SA research, much of it used to develop aircraft training programs, and currently available for cybersecurity training development (Table 12.7).

Table 12.5 Critical Security Controls (CSCs).

Critical Security Controls (CSCs)	Description
1	Inventory of authorized and unauthorized devices
2	Inventory of authorized and unauthorized software
3	Secure configurations for hardware and software on mobile device, laptops, workstations, and servers
4	Continuous vulnerability assessment and remediation
5	Controlled use of administrative privileges
6	Maintenance, monitoring, and analysis of audit logs
7	E-mail and web browser protections
8	Malware defenses
9	Limitation and control of network ports
10	Data recovery capability
11	Secure configurations for network devices
12	Boundary defense
13	Data protection
14	Controlled access based on the need to know
15	Wireless access control
16	Account monitoring and control
17	Security skills assessment and appropriate training to fill gaps
18	Application software security
19	Incident response and management
20	Penetration tests and red team exercises

Table 12.7's methods for evaluating SA are used in conjunction with training systems, providing the simulated cyber terrain for rehearsing known and hypothesized scenarios.

12.2 Cyber Training Systems

One driver of interest in cyber M&S is military applications, where uses of cyber are increasing, as well, with notable uses in Estonia and Georgia over the last decade. While there is work currently being done to characterize cyber systems and their threats, the objective here (Table 12.8) is to look at trainers/simulators for cyber phenomena.

Table 12.6 Australian Signals Directorate computer network defense controls.

Australian Signals Directorate (ASD) Control Name	Description
Application Whitelisting (1)	Whitelisting – when implemented correctly – makes it harder for an adversary to compromise an organization's computer system. Application whitelisting is a technical measure that only allows authorized applications to run on a system. This helps prevent malicious software and unauthorized applications from running.
Patching Systems (2,3)	A software patch is a small piece of software designed to fix problems or update a computer program. Patching an organization's system encompasses both the second and third mitigation strategies. It is important to patch both your operating system and applications within a two-day timeframe for serious vulnerabilities. Once a vulnerability in an operating system or application is made public, you can expect malware to be developed by adversaries within 48 h. In some cases, malware has been developed to take advantage of a publicly disclosed vulnerability within eight hours. There is often a perception that by patching a system without rigorous testing, something is likely to break on the system. In the majority of cases, patching will not affect the function of an organization's computer system. Balancing the risk between taking weeks to test patches and patching serious vulnerabilities within a two-day timeframe can be the difference between a compromised and a protected system.
Restricting Administrative Privileges (4)	When an adversary targets a system, they will primarily look for user accounts with administrative privileges. Administrators are targeted because they have a high level of access to an organization's computer network. If an adversary gains access to a user account with administrative privileges, they can access any data the administrator can access – which generally means everything. Minimizing administrative privileges makes it more difficult for the adversary to spread or hide their existence on a system. Administrative privileges should be tightly controlled. It is important that only staff and contractors that need administrative privileges have them. In these cases, separate accounts with administrative privileges should be created that do not have access to the Internet. This reduces the likelihood of malware infecting the administrator as they should not be web browsing or checking emails while using their privileged account.

Table 12.7 Methods of measuring situational awareness.

Situational awareness measure	Objective
Situational Awareness Global Assessment Technique (SAGAT)	SAGAT is a global tool developed to assess Situation Awareness (SA) across all of its elements based on a comprehensive assessment of operator SA requirements (Endsley 1995) that includes a three-layer model: • Level 1 – the perception of task relevant elements in the environment • Level 2 – the comprehension of their meaning in relation to task goals • Level 3 – the projection of their future states.
Human Potential Explorer (HUPEX)	Culture-independent PC tool for measuring SA under stress
WOMBAT	The WOMBAT Situational Awareness and Stress Tolerance Test is a modern psychological assessment tool for selecting complex-system operators such as pilots, air traffic controllers, ship and train operators, 9-1-1 dispatchers, and nuclear-plant operators; in fact anyone in charge of complex operations involving multiple concurrent inputs and response alternatives.
Situational Awareness Rating Technique (SART) (Taylor et al. 2000)	SART uses the following 10 dimensions to measure operator SA: • familiarity of the situation • focusing of attention • information quantity/quality • instability of the situation • concentration of attention • complexity of the situation • variability of the situation • arousal • spare mental capacity SART is administered post-trial and involves the participant rating each dimension on a seven-point rating scale (1 = Low, 7 = High) in order to gain a subjective measure of SA.

As shown in Table 12.8, multiple training simulations exist for cyber. Industrial control systems (ICS') (Carr 2014), just one example of an enterprise attack surface, includes multiple elements (Javate 2014) for possible use of M&S to train and protect. These tools provide an important contextual view for evaluating a team or individual's SA.

Supplier	Offering	System description
APMG International	Cyber Defense Capability Assessment Tool	This tool links security, IT risk management, and business resilience areas for assessing and enhancing cyber capability of organizations. A software-based framework, the Cyber Defense Capability Assessment Tool models cyber capabilities of an enterprise.
Antycip/Scalable	Network Defense Trainer	Representation of a cyberattack in mission rehearsal scenarios enabling users to identify the main impact(s) on a scenario; uses Exata to emulate the wireless network.
Belden	TOFINO SCADA Security Simulator	Tofino SCADA Security Simulator was a complete SCADA system sold as a portable platform (discontinued)
Boeing	CRIAB/ Cyber-Range-In-A-Box	Cyber Range-In-A-Box (CRIAB) is a compact system used to support the development, test, experimentation, and training of cyber tools and techniques. CRIAB creates security solutions by allowing modeling and simulation of complex missions and advanced threats. CRIAB is Boeing's hardware and software solution for efficient network emulation, virtualization, and integration for training, platform validation, rehearsals, and evaluations. CRIAB is the leading virtual cyber range solution supporting the development and test of tools and techniques, and the training of today's cybersecurity workforce.
Circadence (gaming/training)		Offer an immersive, AI-powered, patent-pending, proprietary cybersecurity training platform
Camber	CENTS/SLAM-R/O&T	The Camber product is the result of an US Air Force initiative, started in 2003 and resulted in what is now called the Air Force Simulator Training and Exercises (SIMTEX) program. CENTS provides the baseline for the HOTSIM (Hands On Training SIMulator) for training individuals and CYNTRS (Cyber Security Network Training Simulator) for training network teams. Components in these simulators are SLAM-R (Sentinel-legion-Autobuild-Myrmidon-Reconstitution) and the RGI (Range Global Internet).
Cybersponse		Cyber responder training; initial critical asset evaluation for security strategy development.
Diatteam	Hynesim (HYbrid NEtwork SIMulator)	The product centralizes around a scenario development tool that providing means to quickly design the environment under test by using Graphical editors, leveraging an extensible set of libraries that can provide the basic blocks of a network. It contains readymade images of different windows – and Linux OS's and allows to create new ones with various patch levels and their vulnerabilities. Also images exist for mainstream CISCO routers and generic images for switches. Once the topology is defined, attributes can be added that control the network in terms of speed, ports been opened, etc.

(Continued)

Table 12.8 (Continued)

Supplier	Offering	System description
Elbit	Cybershield NCDS Training System	The system features an advanced training management system, where the training manager defines, builds, deploys, and runs the training methodology and scenario for each training session. The trainees' activities are tracked and recorded – along with all logs from the network components and security information events – to be fully analyzed during a debriefing and after action review (AAR).
Metova CyberCENTS		Range, attack traffic generation, and training.
Naval Postgraduate School	Malicious Activity Simulation Tool (MAST)	Support the conduct of network administrator security training on the very network that the administrator is supposed to manage. A key element of MAST (Littlejohn and Makhlouf 2013) is to use malware mimics to simulate malware behavior. Malware mimics look and behave like real malware except for the damage that real malware causes.
	CyberSIEGE	CyberCIEGE enhances information assurance and cybersecurity education and training through the use of computer gaming
SANS	Cyber City	NetWars CyberCity is designed to teach warriors and infosec pros that cyber action can have significant kinetic impact in the physical world. CyberCity is a 1:87 scale miniaturized physical city that features SCADA-controlled electrical power distribution, as well as water, transit, hospital, bank, retail, and residential infrastructures. CyberCity engages participants to defend the city's components from terrorist cyberattacks, as well as to utilize offensive tactics to retake or maintain control of critical assets.
Scalable Networks	Network Defence Trainer	The Network Defence Trainer (NDTrainer) is a live-virtual-constructive (LVC) system for implementing cyber-range environments used to train cyber warriors. The NDTrainer system leverages a virtual network model that simulates communication networks. Both live and virtual hosts can be connected to the virtual network model, and the system can be federated with other training simulators to create training solutions.
Selex ES	NCSE	Communications-focused representation of a cyberattack, enabling users to identify where might be the main impact(s) on a scenario
Tele-communications Systems (TCS)	TCS' Art of Exploitation® (AoE™) Portfolio provides that protection with hands-on training and services from trusted and credentialed professionals.	

12.2.1 Scalable Network Defense Trainer (NDT)

This is one of the few tools identified to date that provides a representation of the impact of an event in cyberspace on both the informational and operational capabilities of a mission. It creates linkages between the cyber training environment and the classical domain training exercises. It is service oriented and Computer Generated Forces (CGF)-agnostic adding cyber effects to traditional training effects, thereby training operators to work round a cyberattack and complete their mission objectives. The tool is interoperable with other simulations via HLA to create an emulated software virtual network running in real time.

The NDT cyber tools and ranges provide an engine for representing a cyberattack, and this engine uses the network protocols and standards appropriate to these tools. A DIS/HLA gateway provides an interface to training simulators and simulations, allowing the NDT to deliver a representation of cyber traffic to the federation running the training simulation, which uses the standards appropriate to its own level.

12.2.2 SELEX ES NetComm Simulation Environment (NCSE)

This tool allows users to model and simulate operational network assets. By implementing a "System-in-the-Loop" capability users can establish a "Live-Constructive" connection and allow real hardware or applications and the simulation environment to interact as a common operational picture. It incorporates communications effects with the rest of the simulation to generate an enhanced awareness of the impact a cyberattack might have on the scenario. It can be integrated with "most common CGFs" and using HLA can be federated into a synthetic environment to allow decision makers to understand how a cyberattack might alter the interactions of entities within the scenario.

The NCSE environment is designed to be integrated with virtual training tools to provide cyber personnel with training on communications assets. This allows them to analyze the scalability, survivability, availability, and reliability of the networks. In turn, this leads to an improved SA by enabling users to identify where the main impact of a cyberattack might be noticed and to make appropriate provision. At the network level, the provision could be to ensure a robust configuration that includes an appropriate level of deliberate redundancy. For non-cyber operators, this could be to ensure they are trained to operate in reversionary modes should certain equipment become unavailable.

Both the Scalable NDT and SELEX' NCSE are training platforms to improve the SA of the respective cyber and mission operators. These training platforms are usually targeted toward training mission operators to know when to call the cyber professionals, who maintain specialized tools.

12.2.3 Example Cyber Tool Companies

While it is a challenge to monitor the almost daily evolution of cyber offerings, Table 12.9 provides a sample of cyber-specific companies.

Table 12.9's offerings span from foundational research (e.g. Virginia Tech) to specialized tools for assessing (e.g. Lumeta IPSonar), evaluating (e.g. GNS3), and performing strategy evaluation (e.g. CAULDRON) on a network of interest. While Table 12.9's offerings are already on the market, with a developing user community, looking at recent patents, and patent applications, provide a view of what to expect, in terms of capabilities, over the next few years.

12.3 Cyber-Related Patents and Applications

Granted patents (Table 12.10) provide a view as to what the US Patent and Trademark Office has determined to be a novel contribution to cybersecurity.

Table 12.9 Sample commercial training companies and offerings.

Company	Description
Bivio Networks	The platform deployed for the exercise is part of the Bivio FlowIntelligence application suite that combines the Suricata Engine from OISF, an Open Source Next Generation Intrusion Detection and Prevention Engine, with Symantec Cyber Security: DeepSight™ Intelligence data feeds and the Proofpoint ET Pro Ruleset.
CyVision	CAULDRON leverages Topological Vulnerability Analysis (TVA) approach. TVA monitors the state of network assets, maintains models of network vulnerabilities and residual risk, and combines these to produce models that convey the impact of individual and combined vulnerabilities on the overall security posture. The core element of this tool is an attack graph showing all possible ways an attacker can penetrate the network.
GNS3	Network software emulator for combining real and emulated devices.
Lumeta	IPSonar used for mapping enterprise-level networks.
Neo Technology	Open source graph database – capability for attack graph enumeration.
Rivera Group	EAGLE6 works by automatically building an enterprise model through logs and code repositories.
Virginia Tech Hume Center	The Hume Center leads Virginia Tech's research, education, and outreach programs focused on the challenges of cybersecurity and autonomy in the context of national and homeland security. Education programs provide mentorship, internships, scholarships, and seek to address key challenges in qualified US citizens entering federal service. Current research initiatives include cyber–physical system security, orchestrated missions, and the convergence of cyber warfare and electronic warfare.

Table 12.10 Granted patents.

Patent number	Title	Assignee	Link
9,778,628	Optimization of human supervisors and cyber–physical systems	Goodrich Corporation (Charlotte, NC)	http://patft.uspto.gov/netacgi/nph-Parser?Sect1=PTO2&Sect2=HITOFF&u=%2Fnetahtml%2FPTO%2Fsearch-adv.htm&r=1&p=1&f=G&l=50&d=PTXT&S1=(cyber.TI.+AND+((state+AND+estimator).BSUM.+or+(state+AND+estimator).DETD.+or+(state+AND+estimator).DRWD.))&OS=Ttl/(cyber)+and+Spec/(state+and+estimator)&RS=(TTL/cyber+AND+SPEC/(state+AND+estimator))
9,699,209	Cyber vulnerability scan analyses with actionable feedback	Cyence Inc. (San Mateo, CA)	http://patft.uspto.gov/netacgi/nph-Parser?Sect1=PTO2&Sect2=HITOFF&u=%2Fnetahtml%2FPTO%2Fsearch-adv.htm&r=2&p=1&f=G&l=50&d=PTXT&S1=(cyber.TI.+AND+((state+AND+estimator).BSUM.+or+(state+AND+estimator).DETD.+or+(state+AND+estimator).DRWD.))&OS=Ttl/(cyber)+and+Spec/(state+and+estimator)&RS=(TTL/cyber+AND+SPEC/(state+AND+estimator))
9,680,855	Probabilistic model for cyber risk forecasting	Neo Prime, LLC	http://patft.uspto.gov/netacgi/nph-Parser?Sect1=PTO2&Sect2=HITOFF&u=%2Fnetahtml%2FPTO%2Fsearch-adv.htm&r=1&p=1&f=G&l=50&d=PTXT&S1=((cyber+AND+model).TI.+AND+((state+AND+estimator).BSUM.+or+(state+AND+estimator).DETD.+or+(state+AND+estimator).DRWD.))&OS=Ttl/(cyber+and+model)+and+Spec/(state+and+estimator)&RS=(TTL/(cyber+AND+model)+AND+SPEC/(state+AND+estimator))
9,521,160	Inferential analysis using feedback for extracting and combining cyber risk information	Cyence Inc. (San Mateo, CA)	http://patft.uspto.gov/netacgi/nph-Parser?Sect1=PTO2&Sect2=HITOFF&u=%2Fnetahtml%2FPTO%2Fsearch-adv.htm&r=5&p=1&f=G&l=50&d=PTXT&S1=(cyber.TI.+AND+((state+AND+estimator).BSUM.+or+(state+AND+estimator).DETD.+or+(state+AND+estimator).DRWD.))&OS=Ttl/(cyber)+and+Spec/(state+and+estimator)&RS=(TTL/cyber+AND+SPEC/(state+AND+estimator))
9,258,321	Automated internet threat detection and mitigation system and associated methods	Raytheon Foreground Security, Inc. (Heathrow, FL)	http://patft.uspto.gov/netacgi/nph-Parser?Sect1=PTO2&Sect2=HITOFF&u=%2Fnetahtml%2FPTO%2Fsearch-adv.htm&r=1&p=1&f=G&l=50&d=PTXT&S1=%28cyber+AND+risk%29.ABTX.&OS=abst/%28cyber+and+risk%29&RS=ABST/%28cyber+AND+risk%29
9,253,203	Diversity analysis with actionable feedback methodologies (insurance application)	Cyence Inc. (San Mateo, CA)	http://patft.uspto.gov/netacgi/nph-Parser?Sect1=PTO2&Sect2=HITOFF&u=%2Fnetahtml%2FPTO%2Fsearch-adv.htm&r=2&p=1&f=G&l=50&d=PTXT&S1=%28cyber+AND+risk%29.ABTX.&OS=abst/%28cyber+and+risk%29&RS=ABST/%28cyber+AND+risk%29

(Continued)

Table 12.10 (Continued)

Patent number	Title	Assignee	Link
9,241,008	System, method, and software for cyber threat analysis	Raytheon	http://patft.uspto.gov/netacgi/nph-Parser?Sect1=PTO2&Sect2=HITOFF&u=%2Fnetahtml%2FPTO%2Fsearch-adv.htm&r=1&p=1&f=G&l=50&d=PTXT&S1=9241008.PN.&OS=pn/(9241008)&RS=PN/9241008
9,210,185	Cyber threat monitor and control apparatuses, methods and systems (threat intelligence)	Lookingglass Cyber Solutions, Inc. (Baltimore,MD)	http://patft.uspto.gov/netacgi/nph-Parser?Sect1=PTO2&Sect2=HITOFF&u=%2Fnetahtml%2FPTO%2Fsearch-adv.htm&r=3&p=1&f=G&l=50&d=PTXT&S1=%28cyber+AND+risk%29.ABTX.&OS=abst/%28cyber+and+risk%29&RS=ABST/%28cyber+AND+risk%29
9,177,139	Control system cybersecurity	Honeywell International Inc. (Morristown, NJ)	http://patft.uspto.gov/netacgi/nph-Parser?Sect1=PTO2&Sect2=HITOFF&u=%2Fnetahtml%2FPTO%2Fsearch-adv.htm&r=20&f=G&l=50&d=PTXT&p=1&S1=((((Cyber+AND+Model)+AND+simulation)+AND+State)+AND+estimator)&OS=Cyber+AND+Model+and+simulation+and+State+and+estimator&RS=((((Cyber+AND+Model)+AND+simulation)+AND+State)+AND+estimator)
9,118,714	Apparatuses, methods, and systems for a cyber threat visualization and editing user interface	Lookingglass Cyber Solutions, Inc. (Baltimore, MD)	http://patft.uspto.gov/netacgi/nph-Parser?Sect1=PTO2&Sect2=HITOFF&u=%2Fnetahtml%2FPTO%2Fsearch-adv.htm&r=4&p=1&f=G&l=50&d=PTXT&S1=%28cyber+AND+risk%29.ABTX.&OS=abst/%28cyber+and+risk%29&RS=ABST/%28cyber+AND+risk%29
9,092,631	Computer-implemented security evaluation methods, security evaluation systems, and articles of manufacture	Battelle	http://patft.uspto.gov/netacgi/nph-Parser?Sect1=PTO2&Sect2=HITOFF&u=%2Fnetahtml%2FPTO%2Fsearch-adv.htm&r=1&p=1&f=G&l=50&d=PTXT&S1=9092631.PN.&OS=pn/(9092631)&RS=PN/9092631
8,726,393	Cybersecurity analyzer	ABB Technology AG (Zurich, CH)	http://patft.uspto.gov/netacgi/nph-Parser?Sect1=PTO2&Sect2=HITOFF&u=%2Fnetahtml%2FPTO%2Fsearch-adv.htm&r=6&p=1&f=G&l=50&d=PTXT&S1=%28cyber+AND+risk%29.ABTX.&OS=abst/%28cyber+and+risk%29&RS=ABST/%28cyber+AND+risk%29

8,621,637	Systems, program product, and methods for performing a risk assessment workflow process for plant networks and systems	Saudi Arabian Oil Company (SA)	http://patft.uspto.gov/netacgi/nph-Parser?Sect1=PTO2&Sect2=HITOFF&u=%2Fnetahtml%2FPTO%2Fsearch-adv.htm&r=7&p=1&f=G&l=50&d=PTXT&S1=%28cyber+AND+risk%29.ABTX.&OS=abst/%28cyber+and+risk%29&RS=ABST/%28cyber+AND+risk%29
8,601,587	System, method, and software for cyber threat analysis	Raytheon	http://patft.uspto.gov/netacgi/nph-Parser?Sect1=PTO2&Sect2=HITOFF&u=%2Fnetahtml%2FPTO%2Fsearch-adv.htm&r=1&p=1&f=G&l=50&d=PTXT&S1=8601587.PN.&OS=pn/(8601587)&RS=PN/8601587
8,583,583	Cyber auto tactics techniques and procedures for multiple hypothesis engine	Lockheed Martin	http://patft.uspto.gov/netacgi/nph-Parser?Sect1=PTO2&Sect2=HITOFF&u=%2Fnetahtml%2FPTO%2Fsearch-adv.htm&r=1&p=1&f=G&l=50&d=PTXT&S1=8583583.PN.&OS=pn/(8583583)&RS=PN/8583583

Table 12.11 Patent applications.

Patent application number	Title	Assignee	Link
20160301710	CYBER DEFENSE	CYBERGYM CONTROL LTD	http://appft.uspto.gov/netacgi/nph-Parser?Sect1=PTO2&Sect2=HITOFF&u=%2Fnetahtml%2FPTO%2Fsearch-adv.html&r=2&p=1&f=G&l=50&d=PG01&S1=(Ofir+AND+HASON).IN.&OS=in/(Ofir+and+HASON)&RS=IN/(Ofir+AND+HASON)
20150295948	Method and device for simulating network resilience against attacks	Suzanne Hassell et al.	http://appft.uspto.gov/netacgi/nph-Parser?Sect1=PTO2&Sect2=HITOFF&u=%2Fnetahtml%2FPTO%2Fsearch-adv.html&r=5&p=1&f=G&l=50&d=PG01&S1=(suzanne+AND+hassell).IN.&OS=in/(suzanne+and+hassell)&RS=IN/(suzanne+AND+hassell)
20110288904	System, Method, and Software for Analyzing Maneuvers of an Application in a Distributed Computing Environment	Raytheon	http://appft.uspto.gov/netacgi/nph-Parser?Sect1=PTO2&Sect2=HITOFF&u=%2Fnetahtml%2FPTO%2Fsearch-adv.html&r=7&p=1&f=G&l=50&d=PG01&S1=(suzanne+AND+hassell).IN.&OS=in/(suzanne+and+hassell)&RS=IN/(suzanne+AND+hassell)
20110185432	Cyber Attack Analysis	Raytheon	http://appft.uspto.gov/netacgi/nph-Parser?Sect1=PTO2&Sect2=HITOFF&u=%2Fnetahtml%2FPTO%2Fsearch-adv.html&r=8&p=1&f=G&l=50&d=PG01&S1=(suzanne+AND+hassell).IN.&OS=in/(suzanne+and+hassell)&RS=IN/(suzanne+AND+hassell)
20150106941	Computer-Implemented Security Evaluation Methods, Security Evaluation Systems, and Articles of Manufacture	Batelle	http://appft.uspto.gov/netacgi/nph-Parser?Sect1=PTO2&Sect2=HITOFF&u=%2Fnetahtml%2FPTO%2Fsearch-adv.html&r=1&p=1&f=G&l=50&d=PG01&S1=20150106941.PGNR.&OS=dn/(20150106941)&RS=DN/20150106941

20140245449	SYSTEM, METHOD, AND SOFTWARE FOR CYBER THREAT ANALYSIS	Raytheon	http://appft.uspto.gov/netacgi/nph-Parser?Sect1=PTO2&Sect2=HITOFF&u=%2Fnetahtml%2FPTO%2Fsearch-adv.html&r=1&f=G&l=50&d=PG01&p=1&S1=20140245449.PGNR.&OS=dn/(20140245449)&RS=DN/20140245449
20130347116	THREAT EVALUATION SYSTEM AND METHOD	Zuclu Research LLC	http://appft.uspto.gov/netacgi/nph-Parser?Sect1=PTO2&Sect2=HITOFF&u=%2Fnetahtml%2FPTO%2Fsearch-adv.html&r=1&f=G&l=50&d=PG01&p=1&S1=20130347116.PGNR.&OS=dn/(20130347116)&RS=DN/20130347116
20130055404	System And Method For Providing Impact Modeling And Prediction Of Attacks On Cyber Targets	Aram Khalili	http://appft.uspto.gov/netacgi/nph-Parser?Sect1=PTO2&Sect2=HITOFF&u=%2Fnetahtml%2FPTO%2Fsearch-adv.html&r=1&f=G&l=50&d=PG01&p=1&S1=20130055404.PGNR.&OS=dn/(20130055404)&RS=DN/20130055404
20090326899	SYSTEM AND METHOD FOR SIMULATING NETWORK ATTACKS	Q1 Labs	http://appft.uspto.gov/netacgi/nph-Parser?Sect1=PTO2&Sect2=HITOFF&p=1&u=%2Fnetahtml%2FPTO%2Fsearch-adv.html&r=1&f=G&l=50&d=PG01&S1=20090326899.PGNR.&OS=dn/(20090326899)&RS=DN/20090326899
WO 2014066500	CYBER ANALYSIS MODELING EVALUATION FOR OPERATIONS (CAMEO) SIMULATION SYSTEM	Raytheon	https://encrypted.google.com/patents/WO2014066500A1?cl=und
WO 2006121751	Method and system for generating synthetic digital network traffic	Battelle	http://www.google.sr/patents/WO2006121751A1?cl=en

While Table 12.10 shows what is currently protected, in terms of cybersecurity, Table 12.11's innovations are more recent, due to their being applications currently under consideration.

Table 12.11's look at patent applications is constantly evolving, with new applications coming in daily for cyber innovations.

12.4 Conclusions

Table 12.10 and Table 12.11's recent patents and applications provide a glimpse of the innovative activity currently being applied to cyber. M&S, a necessary underpinning for these new inventions, is growing as well. From the time (Table 12.3) or lessons learned (Table 12.5) abstractions that we take from physical security and information assurance, respectively, to the evolving threat data (Table 12.4), the current cyber environment (Table 12.2) provides multiple opportunities for M&S to contribute.

Bibliography

Aaron Pendergrass, J., Lee, S.C., and Durward McDonell, C. (2013). Theory and practice of mechanized software analysis. *Johns Hopkins APL Technical Digest* 32 (2).

Abraham, S. and Nair, S. (2017). Comparative analysis and patch optimization using cyber security analytics framework. *Journal of Defense Modeling and Simulation* 15 (2): 161–180.

Amina, L. (2012). Patent No. 20120096549.

Arnwine, M. (2015). Developing the infrastructure and methodologies for cyber security. In: *18th Annual Systems Engineering Conference*. NDIA.

Australian Government Department of Defence (n.d.). Strategies to Mitigate Cyber Security Incidents. https://www.asd.gov.au/infosec/mitigationstrategies.htm (accessed 4 November 2016).

Backhaus, S., Bent, R., Bono, J. et al. (2013). Cyber physical security: a game theory model of humans interadcting over control systems. *IEEE Transactions on Smart Grid* 4 (4): 2320–2327.

Bakke, C.P. and Suresh K. Damodaran (2015). *The Cyber-Range Event Process*. Cambridge: Cyber Range Interoperability Standards Working Group (CRIS WG).

Becker, J., Knackstedt, R., and Pöppelbuß, J. (2009). Developing maturity models for it management – a procedure model and its application. *Business and Information Systems Engineering* 1 (3): 213–222.

Beidleman, S. (2009). *Defining and Deterring Cyber War*. Carlisle: US Army War College.

Ben-Asher, N., Oltramari, A., Erbacher, R.F., and Gonzalez, C. (2015). Ontology based adaptive systems of cyber defense. *Proceedings of the 10th Internatioanl Conference on Semantic Technology for Intelligence, Defense, and Security (STIDS)*, Fairfax, Virginia (18 November 2015).

Bernier, M. (2015). *Cyber Effects Categorization – The MACE Taxonomy*. TTCP JSA TP3 Cyber Analysis. Ottawa ON: DRDC Center for Operational Research and Analysis.

An Introduction to Cyber Modeling and Simulation, First Edition. Jerry M. Couretas.

Black, F. and Scholes, M. (1973). The pricing of options and corporate liabilities. *The Journal of Political Economy* 81 (3): 637–654.
Bloom, B.S. (1994). Reflections on the development and use of the taxonomy. In: *Bloom's Taxonomy: A Forty-year Retrospective* (ed. K.J. Rehage, L.W. Anderson and L.A. Sosniak). Chicago: National Society for the Study of Education.
Bodeau, D.J., Graubart, R., and Picciotto, J. (2011). *Cyber Resiliency Engineering Framework*. MITRE. MITRE. McLean.
Bohme, R. and Schwartz, G. (2010). Modeling cyber insurance: towards a unifying framework. In: *Workshop on the Economics of Information Security (WEIS)*. Cambridge, MA: Harvard University.
Boyd, J. R. (n.d.). The Essence of Winning and Losing. http://www.danford.net/boyd/essence.htm (accessed 10 February 2018).
Bucher, N. (2012). Simulation and emulation in support of operational networks: "ALWAYS ON". In: *NDIA 15th Annual Systems Engineering Conference*. Washington: NDIA.
Butts, J., Rice, M., and Shenoi, S. (2012). An adversarial model for expressing attacks on control protocols. *Journal of Defense Modeling and Simulation* 9 (3).
Callahan, C.J. (2013). *Security Information and Event Management Tools and Insider Threat Detection*. Monterrey: Naval Postgraduate School.
Cam, H. (2015). Risk assessment by dynamic representation of vulnerability, exploitation, and impact. In: *Cyber Sensing 2015*. Baltimore: SPIE.
Carr, N.B. (2014). *Development of a Tailored Methodology and Forensic Toolkit for Industrial Control Systems Incident Response*. Monterrey: Naval Postgraduate School.
Chadha, R., Bowen, T., and Chiang, C.J., et al. (2016). CyberVAN: a cyber security virtual assured network testbed. *Military Communications Conference, MILCOM 2016*, Baltimore, Maryland, USA. IEEE (1–3 November 2016).
Chapman, I.M., Leblanc, S.P., and Partington, A. (2011). Taxonomy of cyber attacks and simulation of their effects. In: *Proceedings of the 2011 Military Modeling and Simulation Symposium*. San Diego: SCS.
Cheng, C., Tay, W.P., and Huang, G.B. (2012). Extreme learning machines for intrusion detection. In: *The 2012 International Joint Conference on Neural Networks (IJCNN)*, 1–8. Brisbane, Australia: IEEE.
Chi, S.D., Park, J.S., and Lee, J. (2003). A role of DEVS simulation for information assurance. *Conference: Information Security Applications, 4th International Workshop*, WISA, Jeju Island, Korea (25–27 August 2003).
Cho, J.H. and Gao, J. (2016). Cyber war game in temporal networks. *PLoS One* 11 (2).
Cho, J.H., Cam, H., and Oltramari, A. (2016). Effect of personality traits on trust and risk to phishing vulnerability: modeling and analysis. In: *Cognitive Methods in Situation Awareness and Decision Support (CogSIMA), IEEE International Multi-Disciplinary Conference*. Beirut: IEEE.
Choo, C.S., Ng, E.C., Ang, D., and Chua, C.L. (2008). Data farming in Singapore: a brief history. *INFORMS*, Washington, DC, USA (7–10 December 2008).

Clark, A., Sun, K., Bushnell, L., and Poovendran, R. (2015). A game-theoretic approach to ip address randomization in decoy-based cyber defense. In: *International Conference on Decision and Game Theory for Security*, 3–21. Springer International Publishing.

ClearSky Research Team (2017, March 17). (C. C. Security, Producer). http://www.clearskysec.com/iec/#att123 (accessed 10 February 2018).

CNSSI (2010). *National Information Assurance (IA) Glossary (CNSSI 4009)*. Committee on National Security Systems.

Colbaugh, R. and Glass, K. (2012). *Proactive Defense for Evolving Cyber Threats*. Sandia: Sandia National Labs.

Compton, M.D., Hopkinson, K.M., Peterson, G.L., and Moore, J.T. (2010). Using modeling and simulation to examine the benefits of a network tasking order. *Journal of Defense Modeling and Simulation* 9 (3).

Coolihan, J. and Allen, G. (2012). *LVC Architecture Roadmap Implementation – Results of the First Two*. Orlando, FL: Joint Training Integration and Evaluation Center.

Couretas, J.M. (1998a). *System Entity Structure Enterprise Alternative Evaluator*. Tucson: University of Arizona.

Couretas, J.M. (1998b). *SEAE-SES Enterprise Alternative Evaluator: Design and Implementation of a Manufacturing Enterprise Alternative Evaluation Tool*. Tucson: University of Arizona.

Couretas, J.M. (2014). Model based system engineering (MBSE) applied to program oversight and complex system of systems analysis. In: *NDIA Systems Engineering Conference*. Springfield: NDIA (28–30 October 2014).

Couretas J. (2017). A developing science of cyber security – an opportunity for model based engineering and design. *SIMULTECH 2017*, 27 July 2017.

Cyber Security and Information Assurance Interagency Working Group (CSIA IWG) (2006). *Hard Problems List*. Washington: INFOSEC Research Council.

Damodaran, S.K. and Couretas, J.M. (2015). Cyber modeling & simulation for cyber-range events. In: *SummerSim*, 8. San Diego: SCS.

Damodaran, S.K. and Smith, K. (2015). *CRIS Cyber-Range Lexicon*. Cambridge: Cyber Range Interoperability Standards Working Group.

Dandurand, L. and Serrano, O.S. (2013). Towards improved cyber security information sharing. In: *5th International Conference on Cyber Conflict* (ed. K. Podins, J. Stinissen and M. Maybaum), 16. Tallinn: NATO CCD COE Publications.

Davis, J. and Magrath, S. (2013). *A Survey of Cyber Ranges and Testbeds*. Edinburgh, Australia: Defence Science and Technology Organisation.

Defense Advanced Research Projects Agency (DARPA) (2004). *Real-time Evaluation of Cyber-course of Action (COA) Impact on Performance & Effectiveness*. Rome: Air Force Research Lab.

Defense Science Board (2013). *Resilient Military Systems and the Advanced Cyber Threat*. Washington, DC: Office of the Under Secretary of Defense for Acquisition, Technology and Logistics.

Deming, W.E. (1967). Walter A. Shewhart, 1891–1967. *American Statistician* 21: 39–40.

Deming, W.E. (2010). *Some Theory of Sampling*. New York: Dover.

DeMuth, B. and Scharlat, J. (2012). Modeling & simulation of cyber effects in a degraded environment (ManTech). In: *ITEA 2012 Cyber Conference*, 13. ITEA.

Denil, J. (2013). *Verification and Deployment of Software Intensive Systems: A Multi-Paradigm Modeling Approach*. Antwerp: University of Antwerp.

Department of Homeland Security (DHS) (n.d.). ICS CERT. https://ics-cert.us-cert.gov/Assessments (accessed 10 February 2018).

Dietrich, N., Smith, D.N., and Edwards, D. (2011). Development and the Deployment of Cosage 2.0. *WinterSim*, Phoenix, Arizona, USA (p. 8) (11–14 December 2011).

DoD Operatonal Test and Evaluation (DOT&E) (2015). Cyber Security. Washington, DC. https://www.google.com/url?sa=t&rct=j&q=&esrc=s&source=web&cd=1&ved=0ahUKEwjQlu-d083NAhVGyyYKHbE1BJMQFggeMAA&url=http%3A%2F%2Fwww.dote.osd.mil%2Fpub%2Freports%2Ffy2014%2Fpdf%2Fother%2F2014cybersecurity.pdf&usg=AFQjCNESavr2MbmhDdV60KCXjl1N3ROKWw&cad=rja (accessed 29 June 2016).

DOT&E (2013). Test and Evaluation Resources. http://www.dote.osd.mil/pub/reports/FY2013/pdf/other/2013teresources.pdf (accessed 11 February 2018)

DRDC (Canada) (2013a). *Statement of Work for the ARMOUR TD, v2.1*. Ottawa: DRDC.

DRDC (Canada) (2013b). *System Technical Specification for the ARMOUR TD, v2.1*. Ottawa: DRDC.

DRDC (Canada) (2014a). *Architectural Design Document for the Automated Computer Network Defence (ARMOUR) Technology Demonstrator (TD) Contract*. Ottawa: DRDC.

DRDC (Canada) (2014b). *System Concept of Operations (CONOPS) for the Automated Computer Network Defence (ARMOUR) Technology Demonstration (TD) Contract*. Ottawa: DRDC.

DRDC Cyber Defence S&T Program (2012). *An Overview*. Toronto: DRDC.

Duvenage, P. and von Solms, S. (2013). The case for cyber counterintelligence. In: *2013 International Conference on Adaptive Science and Technology (ICAST)*. Pretoria: IEEE.

Endsley, M. (1995). Toward a theory of situation awareness in dynamic systems. *Human Factors Journal* 37 (1): 32–64.

Estefan, J.A. (2008). *Survey of Model-Based Systems Engineering (MBSE) Methodologies*. Seattle: International Council on Systems Engineering (INCOSE).

Feller, W. (1968). *An Introduction to Probability Theory and its Applications*. New York: Wiley.

Ferguson, C. (2014). Distributed cyber T&E. *NDIA Annual T&E Conference* (p. 26). Washington.

FireEye (2017). M-Trends 2017. https://www.fireeye.com/ppc/m-trends-2017.html?utm_source=google&utm_medium=cpc&utm_content=paid-search&gclid=Cj0KCQjw-uzVBRDkARIsALkZAdniLMfO9X-z1aSqYzJsuRVHLVFjroaLajoLjFaTV15jnzjdyyWEvNMaAt5sEALw_wcB (accessed 28 March 2018).

Frankel, M., Scouras, J., and De Simone, A. (2015). *Assessing the Risk of Catastrophic Cyber Attack Lessons from the Electromagnetic Pulse Commission*. Baltimore: Johns Hopkins University Applied Physics Laboratory.

Frei, S., Fiedler, U., and May, M. (2006). Why to adopt a security metric? *Quality of Protection. Advances in Information Security* 23: 1–12.

Friedenthal, S., Moore, A., and Steiner, R. (2011). *A Practical Guide to SysML, Second Edition: The Systems Modeling Language*. Washington: OMG Press.

Friedenthal, S., Moore, A., and Steiner, R. (2012). *A Practical Guide to SysML*. Waltham, MA: Morgan Kaufmann.

Gagnon, M.N., Truelove, J., Kapadia, A. et al. (2010). Net Centric Survivability for Ballistic Missile Defense. In: *First International Symposium on Architecting Critical Systems* (ed. H. Giese), 125–141. Springer.

Gallagher, M. and Horta, M. (2013). Cyber joint munitions effectiveness manual (JMEM). *M&S Journal* 5–14.

Gelbstein, E. (2013). Quantifying information risk and security. *ISACA Journal* 4.

Gollmann, D., Massacci, F., and Artsiom, Y. (2006). *Quality of Protection – Security Measurements and Metrics*. Springer.

Grange, F. and Deiotte, R. (2015). *DEVS Extensions for Uncertainty Quantification Architectures*. Denver: ISSAC Corp.

Grimaila, M., Myers, J., Mills, R.F., and Peterson, G. (2012). Design and analysis of a dynamically configured log-based distributed security event detection methodology. *The Journal of Defense Modeling and Simulation: Applications, Methodology, Technology* 9 (3), 219–241.

Guo, R.J. and Sprague, K. (2016). Replication of human operators' situation assessment and decision making for simulated area reconnaissance in wargames. *Journal of Defense Modeling and Simulation* 13 (2): 213–225.

Guruprasad, S., Ricci, R., and Lepreau, J. (2005). *Integrated Network Experimentation using Simulation and Emulation. Testbeds and Research Infrastructures for the Development of Networks and Communities, Tridentcom*, 204–212. IEEE.

Hamilton, J. (2013). Architecture-based network simulation for cyber security. In: *Winter Simulation Conference*. San Diego: SCS.

Hansen, A.P. (2008). *Cyber Flag – A Realistic Cyberspace Training Construct*. Wright Patterson Air Force Base: AFIT.

Hariri, S., Guangzhi, Q.U., and Dharmagadda, T. (2003). Impact Analysis of Faults and Attacks in Large Scale Networks. *IEEE Security and Privacy* 49–54.

Heckman, K.E., Stech, F.J., Schmoker, B.S., and Thomas, R.K. (2015). Denial and deception in cyber defense. *Computer* 48 (4): 36–44.

Henninger, A. (2008a). *Live Virtual Constructive Architecture Roadmap (LVCAR) Final Report (1 of 5)*. Alexandria, VA: M&S CO Project No. 06OC-TR-001.

Henninger, A. (2008b). *Live Virtual Constructive Architecture Roadmap (LVCAR) Interim Report*. Alexandria, VA: M&SCO.

Henry, V. (2002). *The Compstat Paradigm: Management Accountability in Policing, Business and the Public Sector*. Looseleaf Law Publications.

Hoffman, D. (2010). *The Dead Hand: The Untold Story of the Cold War Arms Race and Its Dangerous Legacy*. Anchor.

Hubbard, D.W., Seiersen, R., and Geer, D.E. (2016). *How to Measure Anything in Cybersecurity Risk*. New York: Wiley.

Humphrey, W. (1989). *Managing the Software Process*. Addison Wesley.

IEEE Std 1278 Series (n.d.). IEEE Standards for Distributed Interactive Simulation (DIS). https://standards.ieee.org/findstds/standard/1278.1-2012.html (accessed 18 February 2018).

IEEE Std 1516 (n.d.). High Level Architecture for M&S (HLA). https://standards.ieee.org/findstds/standard/1516-2010.html (accessed 18 February 2018).

IEEE Std 1730-2010 (n.d.). IEEE Recommended Practice for Distributed Simulation Engineering and Execution Process (DSEEP). https://standards.ieee.org/findstds/standard/1730-2010.html (accessed 18 February 2018).

Ingols, K., Lippmann, R., and Piwowarski, K. (2006). Practical attack graph generation for network defense. In: *22nd Annual Computer Security Applications Conference (ACSAC)*. IEEE.

Ivers, J. (2017, 30 March). Security Week. Security vs. Quality: What's the Difference? https://www.securityweek.com/security-vs-quality-what%E2%80%99s-difference (accessed 10 February 2018).

Jabbour, K. and Poisson, J. (2016). Cyber risk assessment in distributed information systems. *The Cyber Defense Review* 1 (1).

Jajodia, S., Shakarian, P., Subrahmanian, V.S. et al. (2015). *Cyber Warfare: Building the Scientific Foundation*. Springer.

Jaquith, A. (2007). *Security Metrics: Replacing Fear, Uncertainty, and Doubt*. Addison-Wesley, Pearson Education.

Javate, M.S. (2014). *Study of Adversarial and defensive components in an Experimental Machinery Control Systems Laboratory Environment*. Monterrey: Naval Postgraduate School.

Jens, P., Niehaves, B., Simons, A., and Becker, J. (2011). Maturity models in information systems research: literature search and analysis. *Communications of the Association for Information Systems* 29 (1).

Johnson, R.E., Isensee, E.K., and Allison, W.T. (1995). A stochastic version of the concepts evaluation model (CEM). *Naval Research Logistics* 233–246.

Joint Chiefs of Staff. (2014). *Information Operations*. Joint Publication 3-13, Joint Chiefs of Staff, Washington.

Jones, J. (2005). *An Introduction to Factor Analysis of Information Risk (FAIR)*. Risk Management Institute.

Jones, R.M., O'Grady, R., and Nicholson, D. (2015). Modeling and integrating cognitive agents within the emerging cyber domain. In: *Proceedings of the*

Interservice / Industry Training, Simulation and Education Conference (I/ITSEC). Orlando: NDIA.

Jonsson, E. and Olovsson, T. (1997). A quantitative model of the security intrusion process based on attacker behavior. *IEEE Transactions on Software Engineering* 23 (4): 235–245.

Kavak, H. (2016). A characterization of cybersecurity simulation scenarios. *Proceedings of the 19th Communications & Networking Symposium.* ACM.

Khintchine, A. (1969). *Mathematical Methods in the Theory of Queueing.* New York: Hafter.

Kick, J. (2014). *Cyber Exercise Playbook.* Wiesbaden, Germany: MITRE.

Kiesling, E., Ekelhart, A., Grill, B. et al. (2013). Simulation-based optimization of information security controls: an adversary-centric approach. In: *Winter Simulations Conference (WSC),* 2054–2065. Washington, DC, USA: IEEE.

Kim, T., Hwang, M.H., and Kim, D. (2008). *DEVS/NS-2 Environment – Integrated Tool for Efficient Networks Modeling and Simulation.* San Diego: SCS.

King, S. (2011). *Cyber S&T Priority Steering Council Research Roadmap.* Washington: NDIA Disruptive Technologies Conference.

Klein, G., Elphinstone, K., and Heiser, G. (2009). seL4: formal verification of an OS kernel. In: *Proceedings of the ACM SIGOPS 22nd Symposium on Operating Systems Principles,* 207–220. New York: ACM.

Kossiakoff, A., Sweet, W.N., Seymour, S.J., and Biemer, S.M. (2011). *Systems Engineering: Principles and Practice.* New York: Wiley.

Kotenko, I. (2005). Agent-based modeling and simulation of cyber-warfare between malefactors and security agents in internet. In: *Proceedings 19th European Conference on Modelling and Simulation* (ed. Y.M.R. Zobel and E. Kerckhoffs).

Kotenko, I. and Chechulin, A. (2013). A cyber attack modeling and impact assessment framework. In: *Proceedings of the 5th International Conference on Cyber Conflict.* New York: IEEE.

Kotenko, I., Konovalov, A., and Shorov, A. (2012). Agent-based simulation of cooperative defence against botnets. *Concurrency Computation Practice and Experience* 573–588.

Kott, A. (2014). Towards fundamental science of cyber security. In: *Network Science and Cybersecurity* (ed. R.E. Pino), 1–13. New York: Springer.

Kott, A., Stoianov, N., Baykal, N. et al. (2015). *Assessing Mission Impact of Cyberattacks: Report of the NATO IST-128 Workshop.* Adelphi: Army Research Lab (ARL-TR-7566).

Krepinevich, A. (2012). *Cyber Warfare: A "Nuclear Option".* Washington: Center for Strategic and Budgetary Assessments (CSBA).

Lange, M., Kott, A., Ben-Asher, N. et al. (2017). *Recommendations for Model-Driven Paradigms for Integrated Approaches to Cyber Defense.* Adelphi: Army Research Lab.

Lavigne, V. and Gouin, D. (2014). Visual analytics for cyber security and intelligence. *Journal of Defense Modeling and Simulation* 11 (2): 175–199.

Lee, R.M., Assante, M.J., and Conway, T. (2014). German Steel Mill Cyber Attack. SANS ICS CP/PE (Cyber-to-Physical or Process Effects) Case Study Paper. SANS.

Leversage, D.J. and Byres, E.J. (2007). Comparing electronic battlefields: using mean time-to-compromise as a comparative security metric. In: *Computer Network Security* (ed. I.K. Vladimir Gordodetsky). New York: Springer.

Lewis, J. (2015). *Deterrence in the Cyber Age*. Washington: Center for Strategic and International Studies.

Li, W. and Vaughn, R. (2006). Cluster security research involving the modeling of network exploitations using exploitation graphs. *Sixth IEEE International Symposium on Cluster Computing and Grid Workshops*, Singapore (30 May 2006).

Littlejohn, A.M. and Makhlouf, E. (2013). *Test and evaluation of the Malicious Activity Simulation Tool (MAST) in a Local Area Network (LAN) running the Common PC Operating System Environment (COMPOSE)*. Monterrey: Naval Postgraduate School.

Lo, A.W. and Hasanhodzic, J. (2010). *The Evolution of Technical Analysis*. Hoboken: Wiley.

Luenberger, D. (1979). *Introduction to Dynamic Systems: Theory, Models and Applications*. New York: Wiley.

Lyons, K. (2014). *A Recommender System in the Cyber Defense Domain*. Wright-Patterson Air Force Base, OH: AFIT.

Malekzadeh, M., Ghani, A.A.A., Subramaniam, S., and Desa, J. (2011). Validating Reliability of OMNeT in Wireless Networks DoS Attacks: Simulation vs. Testbed. *International Journal of Network Security* 13–21.

Manadhata, P. and Wing, J.M. (2008). *An Attack Surface Metric*. Pittsburgh: CMU.

Mandiant (2014). *M-Trends – Beyond the Breach*. FireEye.

Manshaei, M.H., Zhu, Q., Alpcan, T. et al. (2013). Game theory meets network security and privacy. *ACM Computing Surveys* 45 (3).

MANTECH. (2018). Cybersecurity. http://www.mantech.com/solutions/Cyber%20Security/Pages/default.aspx (accessed 11 February 2018).

Marshall, H., Mize, J.R., Hooper, M. et al. (2015). *Cyber Operations Battlefield Web Services (COBWebS) – Concept for a Tactical Cyber Warfare Effect Training Prototype. SIW*, 8. Orlando: SISO.

Masi, D., Fischer, M., Shortle, J.F., and Chen, C.H. (2011). Simulating network cyber attacks using splitting techniques. In: *Proceedings of the Winter Simulation Conference*, 3212–3223. Pheonix: IEEE.

Maynard, T. and Beecroft, N. (2015). *Business Blackout – The insurance implications of a cyber attack on the US power grid*. London: Lloyd's.

McQueen, M.A., Boyer, W.F., Flynn, M.A., and Beitel, G.A. (2006). Time-to-compromise model for cyber risk reduction estimation. In: *Quality of Protection* (ed. D. Golman). New York: Springer.

Millett, L.I., Fischhoff, B., and Weinberger, P.J. (2017). *Foundational Cybersecurity Research: Improving Science, Engineering, and Institutions*. Washington: National Academy of Sciences, Division on Engineering and Physical Sciences.

MITRE (2014, July 3). Cybersecurity. Collaborative Research Into Threats (CRITs). https://www.mitre.org/capabilities/cybersecurity/overview/cybersecurity-blog/collaborative-research-into-threats-crits (accessed 10 February 2018).

MITRE (2015). *An Overview of MITRE Cyber Situational Awareness Solutions*. McLean: MITRE.

MITRE (n.d.-a). https://www.mitre.org/capabilities/cybersecurity/overview/cybersecurity-resources/standards (accessed 10 February 2018).

MITRE (n.d.-b). https://cybox.mitre.org/language/version2.0/ (accessed 10 February 2018).

MITRE (n.d.-c). https://www.mitre.org/publications/technical-papers/standardizing-cyber-threat-intelligence-information-with-the (accessed 10 February 2018).

MITRE (n.d.-d). https://www.google.com/url?sa=t&rct=j&q=&esrc=s&source=web&cd=3&cad=rja&uact=8&ved=0ahUKEwiNsr-i4cjOAhXC2yYKHUSMCFwQFggpMAI&url=https%3A%2F%2Ftaxii.mitre.org%2Fabout%2Fdocuments%2FIntroduction_to_TAXII_White_Paper_November_2012.pdf&usg=AFQjCNESIoOZhB8dpH4 (accessed 10 February 2018).

Morse, K.L., Bryan, D.S., Drake, D.L., and Wells, W.D. (2014a). Realizing the Cyber Operational Architecture Training System (COATS) through standards. In: *SIW*. Orlando: SISO.

Morse, K.L., Drake, D.L., Wells, D., and Bryan, D. (2014b). Realizing the Cyber Operational Architecture Training System (COATS) through standards. In: *SIW*, 10. Orlando: Simulation Interoperability and Standards Organization (SISO).

Musman, S. and Temin, A. (2017). Playing the cyber security game: an approach to cyber security and resilience decision making. *Journal of Defense Modeling and Simulation* 15 (2).

Musman, S., Temin, A., Tanner, M. et al. (2013). Evaluating the impact of cyber attacks on missions. *M&S Journal* 25.

Nadeem, A. and Howarth, M. (2013). Protection of MANETs from a range of attacks using an intrusion detection and prevention system. *Telecommun Systems* 52: 2047–2058.

National Academy of Engineering (1995). Chapter: 3 Integrated Product and Process Design. Information Technology for Manufacturing: A Research Agenda. https://www.nap.edu/read/4815/chapter/5 (accessed 10 February 2018).

National Vulnerability Database (2014). https://nvd.nist.gov/ (accessed 18 February 2018).

NATO (2012). *NATO Allied Joint Doctrine for Information Operations*. Brussels: NATO.

NATO (2014). *AMSP-03 M&S Standards Profile for NATO and Multinational Computer Assisted eXercises with Distributed Simulation*. NATO.

NATO (2015, March). AMSP-01 M&S Standards Profile, Edition C v 1.

NATO Cooperative Cyber Defense Center of Excellence (2018). Exercise Crossed Swords Practised Cyber-Kinetic Operations in Latvia. https://ccdcoe.org/exercise-crossed-swords-practised-cyber-kinetic-operations-latvia.html (accessed 10 February 2018).

NATO MSG 117 (2015, publication pending). *Exploiting Modeling and Simulation in Support of Cyber Defense*. Brussels: NATO.

NEWSWEEK (2016). Alleged dam hacking raises fears of cyber threats to infrastructure (30 March). http://www.newsweek.com/cyber-attack-rye-dam-iran-441940 (accessed 10 February 2018).

Nilsson, N.J. (1998). *Artificial Intelligence*. New York: Morgan Kaufmann.

Noel, S., Ludwig, J., Jain, P., et al. (2015). *Analyzing Mission Impacts of Cyber Actions (AMICA)*.

Norman, R. and Christopher, E.D. (2013). Cyber operations research and network analysis (Corona) enables rapidly reconfigurable cyberspace test and experimentation. *M&S Journal* 15–24.

Norton, C.T. (1979, January). Blue Flag. In: *Air University Review*.

Nunes-Vaz, R., Lord, S., and Ciuk, J. (2011). A More Rigorous Framework for Security-in-Depth. *Journal of Applied Security Researh* 23.

Nunes-Vaz, R., Lord, S., and Bilusich, D. (2014). From strategic security risks to national capability priorities. *Security Challenges* 10 (3): 23–49.

Nutaro, J. (2016). Towards improving software security by using simulation to inform requirements and conceptual design. *Journal of Defense Modeling and Simulation* 13 (1).

Object Management Group (OMG) (2012, June). OMG Systems Modeling Language (OMG SysML™).

Okhravi, H., Rabe, M.A., Mayberry, T.J. et al. (2013b). *Survey of Cyber Moving Target Techniques*. Boston: Lincoln Labs.

Ortalo, R., Deswarte, Y., and Kaâniche, M. (1999). Experimenting with quantitative. *IEEE Transactions on Software Engineering* 25: 633–650.

Panton, M.B.C., Colombi, J.M., Grimaila, M.R., and Mills, R.F. (2014). Strengthening DoD cyber security with the vulnerability market. *Defense ARJ* 21 (1): 466–484.

Park, J.S., Lee, J.S., Kim, H.K., and Chi, S.D. (2001). SECUSIM: a tool for the cyber-attack simulation. *Third International Conference on Information and Communications Security, ICICS 2001*, Xian, China (13–16 November 2001).

Pathmanathan, A. (2013, November 14). *30th Annual International Test and Evaluation Symposium (ITEA)*. http://www.itea.org/~iteaorg/images/pdf/conferences/2013_Annual/Panel_2_Pathmanathan.pdf (accessed 6 January 2015).

Paulenich, J., Agbedo, C., and Rea, K. (2014). *Identification and Triage of Compromised Virtual Machines*. Monterrey, CA, USA: Naval Postgraduate School.

Pawlick, J., Farhang, S., and Quanyan, Z. (2015). Flip the cloud: cyber-physical signaling games in the presence of advanced persistent threats. In: *International Conference on Decision and Game Theory for Security*, 289–308. Springer International Publishing.

Peter Beiling, B.H. (2016). *Methodology for Anticipating and Responding to Successful Cyber Attacks on Physical Systems.* Virginia: Alexandria.

Priest, B.W., Vuksani, E., Wagner, N. et al. (2014). Agent-based simulation in support of moving target cyber defense technology development and evaluation. In: *IEEE Symposium on Security and Privacy*, 12. San Jose: IEEE.

Priest, B.W., Vuksani, E., Wagner, N. et al. (2015a). Agent-based simulation in support of moving target cyber defense technology development and evaluation. In: *SpringSim*, 8. Alexandria: ACM.

Priest, B.W., Vuksani, E., Wagner, N. et al. (2015b). Agent-based simulation in support of moving target cyber defense technology development and evaluation. In: *Proceedings of the 18th Symposium on Communications & Networking*, 16–23. San Diego, USA: Society for Computer Simulation International.

Ratonel, C. (2013). *Cyber Security Simulation Overview Brief.* Metron: Reston.

Raulerson, E.L., Hopkinson, K.M., and Laviers, K.R. (2014). A framework to facilitate cyber defense situational awareness modeled in an emulated virtual machine testbed. *Journal of Defense Modeling and Simulation* 12 (3): 229–239.

Rescorla, E. (2005). Is finding security holes a good idea? *IEEE Security and Privacy* 14–19.

Richards, J.E. (2014). *Using the Department of Defense Architecture Framework to Develop Security Requirements.* SANS: SANS Institute.

Rimondini, M. (2007). *Emulation of Computer Networks with Netkit.* Department of Information Automation, Roma Tre University.

Robinson, D. and Cybenko, G. (2012). A Cyber-based behavioral model. *Journal of Defense Modeling and Simulation* 9 (3).

Romanosky, S. (2016). Examining the costs and causes of cyber incidents. *Journal of Cybersecurity* 2 (2).

Romanosky, S., Ablon, L., Kuehn, A., and Jones, T. (n.d.). Content Analysis of Cyber Insurance Policies: How Do Carriers Write Policies and Price Cybersecurity Risk?. SSRN. https://papers.ssrn.com/sol3/papers.cfm?abstract_id=2929137 (accessed 25 October 2017).

Rossey, L.M., Cunningham, R.K., Fried, D.J. et al. (2002). LARIAT: lincoln adaptable real-time information assurance testbed. In: *Aerospace Conference Proceedings*, vol. 6, 6-2671–2676, 6-2678–6-2682. IEEE.

Rowe, C., Zadeh, H.S., and Garanovich, I.L. (2017). Prioritising investment in military cyber capability using risk analysis. *Journal of Defense Modeling and Simulation.*

Roza, M., Voogd, J., van Emmerik, M., and van Lier, A. (2010). Generic methodology for verification and validation for training simulations.

In: *Interservice/Industry Training, Simulation, and Education Conference (I/ITSEC)*, 12. Orlando: NDIA.

Roza, M., Voogd, J., and Sebalji, D. (2013). The generic methodology for verification and validation to support acceptance of models, simulations and data. *Journal of Defense Modeling and Simulation* 347–365.

Saadawi, H. and Wainer, G. (2013). Principles of discrete event system specification model verification. *Simulation* 41–67.

Saadawi, H., Wainer, G., and Moallemi, M. (2012). Principles of model verification for real-time embedded applications. In: *Real-Time Simulation Technologies: Principles, Methodologies and Applications* (ed. P.M.K. Popovici). Boca Raton, FL: CRC Press.

Sallhammar, K., Helvik, B.E., and Knapskog, S.J. (2006). On stochastic modeling for integrated security and dependability evaluation. *Journal of Networks* 1 (50).

SANS Institute (2006). *A Guide to Security Metrics*. Bethesda: SANS.

Schostack, A. (2014). *Threat Modeling: Designing for Security*. New York: Wiley.

Serban, C., Poylisher, A., Sapello, A. et al. (2015). Testing android devices for tactical networks: a hybrid emulation testbed approach. In: *Proceedings of the Military Communications Conference*. New York: IEEE.

Serdiouk, V. (2007). Technologies for protection against insider attacks on computer systems. In: *Fourth International Conference on Mathematical Methods, Models, and Architectures for Computer Network Security, MMM-ACNS-2007* (ed. V. Gorodetsky, I. Kotenko and V. Skormin), 75–84. St. Petersburg: Springer.

Shakshuki, E.M., Kang, N., and Sheltami, T.R. (2013). EAACK – a secure intrusion-detection system for MANETs. *IEEE Transactions on Industrial Electronics* 60 (3): 1089–1098.

Simonsson, M., Johnson, P., and Wijkström, H. (2007). Model-based IT governance maturity assessments with COBIT. In: *ECIS 2007 Proceedings*. ECIS.

Simulation Interoperability Standards Organization (SISO) (n.d.). Federation Engineering Agreements Template (FEAT). http://www.sisostds.org/FEATProgrammersReference/ (accessed 6 January 2015).

SISO (1999). *Fidelity Implementation Study Group Report*. Orlando: SISO.

SISO-STD-007-2008 (n.d.). Standard for Military Scenario Definition Language.

SISO-STD-011-2014 (n.d.). Standard for Coalition Battle Management Language (C-BML) Phase I.

Skare, P. M. (2013). Patent No. 8595831.

Small Business Innovative Research (SBIR) (2012). Cyber-to-Physical Domain Mapping Toolkit for Vulnerability Analysis and Critical Resource Identification Enablement (CEPHEID VARIABLE). Award Details. https://www.sbir.gov/sbirsearch/detail/393789 (accessed 11 February 2018).

Sommestad, T. (2013). The cybewr security modeling language: a tool for assessing the vulnerability of enterprise system architectures. *IEEE Systems Journal* 7 (3): 363–373.

Stella Croom-Johnson, J.M. (2016). Cyber tools and standards to improve situational awareness. In: *Simulation Interoperability Standards Organization*, 12. Orlando: SISO.

Stine, K. (2012). Inside NIST's cybersecurity strategy. *Washington Technology*.

Streilein, W.W., Truelove, J., Meiners, C.R., and Eakman, G. (2011). Cyber situational awareness through operational streaming analysis. In: *Military Communiations Conference*, 1152–1157. IEEE.

Symantec (2014). *Dragonfly: Cyberespionage Attacks Against Energy Suppliers*.

Taguchi, G., Chowdhury, S., and Wu, Y. (2004). *Taguchi's Quality Engineering Handbook*. New York: Wiley.

Taylor, J.G., Yildirim, U.Z., and Murphy, W.S. (2000). Hierarchy of models approach for aggregated attrition. In: *Winter Simulation Conference*. San Diego: Society for Computer Simulation International.

Tello, B., Winterrose, M., Baah, G., and Zhivich, M. (2015). Simulation based evaluation of a code diversification strategy. *5th International Conference on Simulation and Modeling Methodologies, Technologies and Applications* (pp. 36–43), Colmar, Alsace, France. SIMULTECH 2015.

The DETER Testbed: Overview (2004, August 25). http://www.isi.edu/deter/docs/testbed.overview.pdf (accessed 6 May 2015).

The Ponemon Institute, LLC. (2014). Privileged User Abuse & The Insider Threat. http://www.trustedcs.com/resources/whitepapers/Ponemon-RaytheonPrivilegedUserAbuseResearchReport.pdf (accessed 26 May 2014).

Thompson, M.F. and Irvine, C.E. (2011). Active Learning with the CyberCIEGE Video Game. *4th Workshop on Cyber Security Experimentation and Test*, San Francisco, CA (8–12 August 2011).

Tolk, A. and Muguira, J.A. (2003). The levels of conceptual interoperability model (LCIM). In: *IEEE Fall Simulation Interoperability Workshop*. Orlando: IEEE CS Press.

Torres, G. (2015). *Test & Evaluation/Science & Technology Net-Centric Systems Test (NST) Focus Area Overview*. Pt Mugu: USC, Center for Systems and Software Engineering.

Toutonji, O.A., Yoo, S.M., and Park, M. (2012). Stability analysis of VEISV propagation modeling for network worm attack. *Applied Mathematical Modeling* 2751–2761.

Valilai, O.F. and Houshmand, M. (2009). Advantages of using SysML compatible with ISO 10303-233 for product design and development based on STEP standard. In: *Proceedings of the World Congress on Engineering and Computer Science*, vol. II. San Francisco: WCECS.

Velez, T.U. and Morana, M.M. (2015). *Risk Centric Threat Modeling: process for attack simulation and threat analysis*. Hoboken: John Wiley & Sons, Inc.

Waag, G.L., Kenneth Heist, R., Feinberg, J.M., and Painchaud, L.J. (2001). *Information Assurance Modeling & Simulation (IA M&S) State of the Art Report – A Summary*. Alexandria: MSIAC.

Wagner, N., Lippmann, R., Winterrose, M. et al. (2015). Agent-based simulation for assessing network security risk due to unauthorized hardware. In: *Proceedings of the Symposium on Agent-Directed Simulation*, 18–26. Society for Computer Simulation International.

Wagner, N., Sahin, C.S., Winterrose, M. et al. (2017). Quantifying the mission impact of network-level cyber defensive. *Journal of Defense Modeling and Simulation* 14 (3): 201–216.

Waltz, E. (2000). *Information Warfare: Principles and Operations*. Boston: ArTech House.

Wang, L., Liu, A., and Jajodia, S. (2006). Using attack graphs for correlating, hypothsizing and predicting instrusion alerts. *Computer Communications* 29 (15): 2917–2933.

Wells, D. and Bryan, D. (2015). Cyber operational architecture training system – cyber for all. In: *Interservice/Industry Training, Simulation, and Education Conference (I/ITSEC)*, 9. Orlando: NDIA.

White, G. (2007). The community cyber security maturity model. *Proceedings of the 40th Hawaii International Conference on System Sciences*, Waikoloa, HI, USA (3–6 January 2007).

Wymore, W. (1967a). *1967. A Mathematical Theory of Systems Engineering: The Elements*. Huntington, NY: Krieger.

Wymore, W. (1967b). *A Mathematical Theory of Systems Engineering: The Elements*. New York: Wiley.

Yan, W., Xue, Y., Li, X. et al. (2012). Integrated simulation and emulation platform for cyber-physical system security experimentation. In: *HiCoNS'12*, 8. Beijing: ACM.

Yap, G. (2009). *When is a Hack an Attack? A Sovereign State's Options if Attacked in Cyberspace: A Case Study of Estonia 2007*. Birmingham: Air Command and Staff College Air University Maxwell Air Force Base, Alabama.

Yildrim, U.Z. (1999). *Extending the State-of-the-art for the COMAN/ATCAL Methodology*. Monterrey: NPS.

Yu, S., Gu, G., Barnawi, A. et al. (2015). Malware propagation in large-scale networks. *IEEE Transactions on Knowledge and Data Engineering* 170–179.

Yufik, Y. (2014). Understanding cyber warfare. In: *Network Science and Cybersecurity* (ed. R.E. Pino), 75–91. New York: Springer.

Zeigler, B.P. and Nutaro, J.J. (2016). Towards a framework for more robust validation and verification of simulation models for systems of systems. *Journal of Defense Modeling and Simulation* 3–16.

Zeigler, B.P., Praehofer, H., and Kim, T.G. (2000). *Theory of Modeling and Simulation*. New York: Academic Press.

Zetter, K. (2014). Meet MonsterMind, the NSA Bot That Could Wage Cyberwar Autonomously. *Wired* (13 August).

Index

An Introduction to Cyber Modeling and Simulation, First Edition. Jerry M. Couretas.